RELIGIOUS REPRESENTATION IN PLACE

Religion and Spatial Studies

Series editor: Mark K. George

The critical study and analysis of space examines a wide range of spatial questions, including how religions and religious practitioners build, arrange, and live within physical space; how they conceive and logically organize, classify, and design space; and the complex socio-religious meanings and significations ascribed to space by those who use and occupy it, including meanings that often become the basis for contestation and conflict. This series also analyzes how religion stands alongside various disciplines as opposed to being subsumed by them.

Religious Representation in Place: Exploring Meaningful Spaces at the Intersection of the Humanities and Sciences
Edited by Mark K. George and Daria Pezzoli-Olgiati

Foucault's Heterotopia in Christian Catacombs: Constructing Spaces and Symbols in Ancient Rome
Eric C. Smith

Religious Representation in Place

Exploring Meaningful Spaces at the Intersection of the Humanities and Sciences

Edited by
Mark K. George
and
Daria Pezzoli-Olgiati

palgrave
macmillan

RELIGIOUS REPRESENTATION IN PLACE

First published in 2014 by
PALGRAVE MACMILLAN®
in the United States—a division of St. Martin's Press LLC,
175 Fifth Avenue, New York, NY 10010.

Where this book is distributed in the UK, Europe and the rest of the world, this is by Palgrave Macmillan, a division of Macmillan Publishers Limited, registered in England, company number 785998, of Houndmills, Basingstoke, Hampshire RG21 6XS.

Palgrave Macmillan is the global academic imprint of the above companies and has companies and representatives throughout the world.

Palgrave® and Macmillan® are registered trademarks in the United States, the United Kingdom, Europe and other countries.

ISBN: 978–1–137–37133–1

Library of Congress Cataloging-in-Publication Data

Religious representation in place : exploring meaningful spaces at the intersection of the humanities and sciences / edited by Mark K. George, Daria Pezzoli-Olgiati.
pages cm.—(Religion and spatial studies)
Includes bibliographical references and index.
ISBN 978–1–137–37133–1 (alk. paper)
1. Sacred space. 2. Space—Religious aspects. 3. Place (Philosophy) 4. Humanities. 5. Science. I. George, Mark K., 1962– editor.

BL580.R46 2014
203′.5—dc23 2014008874

A catalogue record of the book is available from the British Library.

Design by Newgen Knowledge Works (P) Ltd., Chennai, India.

First edition: October 2014

10 9 8 7 6 5 4 3 2 1

To

Quinn and Ryan Martindale-George

Noè Archimede and Leandro Elia Pezzoli

Contents

Section II Representing Space in Ancient and Medieval Contexts

Section III Representing Space in Contemporary Contexts

Figures

Preface

The impetus for this project originated in informal conversations between the editors at a Society of Biblical Literature meeting about working together on a shared interest in the intersection of religion and space. In a meeting at the Swiss Institute in Rome 2009, the focus and scope of the project became clear: to hold a conference consisting of an international group of scholars from a wide range of disciplines to explore different interactions of religion and space and then to published revised versions of the conference papers. That conference, "Meaningful Spaces," was held in Zürich at the Collegium Helvetica in April 2011. This book contains revised papers that reflect the conversations, energy, and deepened understandings that characterized this open and engaged academic exchange.

Acknowledgments

This book would not be possible without the help and assistance of a large number of individuals, foundations, and institutions. The editors would like to thank the Swiss National Science Foundation, the Hochschulstiftung of the University of Zurich, the Zürcher Universitätsverein and the Centre for Religion, Economy and Politics (ZRWP), and Iliff School of Theology for their generous financial support of the conference and project. We also thank the Collegium Helveticum for hosting the "Meaningful Spaces" conference in April, 2011. Many thanks are due to the conference and book contributors, all of whom made possible a fascinating and enriching conversation and publication across disciplinary boundaries that typically are not crossed. Their willingness to engage in this project and undertake the work of speaking and listening in English (usually a second, third, or fourth language) to scholars from fields and parts of the world different from their own calls forth our sincere appreciation. Many other people have been involved in the production of this book. We thank Dolores Zoe Bertschinger, Donnie Featherston, and Sean Ryan for the many hours of assistance they provided, always in good humor, with the preparation of the manuscript. Our thanks go to Palgrave Macmillan, our editors there, Burke Gerstenschlager and Brigitte Shull, and their assistants, Madeleine Crum and Caroline Kracunas, for their work, support, and guidance with the book. An additional word of thanks is due to Iliff School of Theology for a quarter leave at the end of this project, which made possible, among other things, the time necessary for final editorial and other work in preparation of the manuscript. We would be remiss if we did not thank the developers of Skype, the voice-over-IP service that greatly facilitated collaboration between the editors. The realities of space and place were constant for us as we engaged in this project, and Internet communication was important for our work. Finally, we thank our families for their patience and support as we pursued this project, whether in Rome, Zürich, San Francisco, or on Skype. They have heard about it for a long time, and we are happy now to have this book to show for it.

Abbreviations

AB	Anchor Bible
ABD	*Anchor Bible Dictionary*
ATD	Das Alte Testament Deutsch
BASOR	*Bulletin of the American Schools of Oriental Research*
BCE	Before Common Era
BCN	Barcelona
BKAT	Biblischer Kommentar. Altes Testament. Edited by M. Noth and H. W. Wolff
BTMC	Bodhgayā Temple Management Committee
BZAW	Beihefte zur Zeitschrift für die alttestamentliche Wissenschaft
cf.	compare
ch(s).	chapter(s)
cm	centimeter
Ed.	Editor
ESA	European Space Agency
Exod	Exodus
Fig.	Figure
Gen	Genesis
HBS	Herders Biblische Studien
HUDCO	Housing and Urban Development Corporation of India
ICC	International Critical Commentary
IGN	Instituto Geográfico Nacional
JBL	*Journal of Biblical Literature*
JSOT	*Journal for the Study of the Old Testament*
Kgs	Kings
km	kilometer
Lev	Leviticus
LHBOTS	Library of Hebrew Bible/Old Testament Studies
NASA	National Aeronautics and Space Administration

NIV	New International Version (Bible)
Nr	Number
NRSV	New Revised Standard Version
NTOA	Novum Testamentum et Orbis Antiquus
Num	Numbers
OBO	Orbis biblicus et orientalis
OTL	Old Testament Library
POV	Point of view
Rev	Revelation
RSV	Revised Standard Version
SBLAIL	Society of Biblical Literature: Ancient Israel and Its Literature
trans.	translator
UNESCO	United Nations Educational, Scientific and Cultural Organization
v.	verse
vol.	volume
WBC	Word Biblical Commentary
WHS	World Heritage Site

Introduction

Mark K. George and Daria Pezzoli-Olgiati

Mont Saint Michel is a small tidal island at the mouth of the Couesnon River along the Normandy coast in northern France. Atop the island, some 240 feet (or about 73 meters) high, is the Abbey of Mont Saint Michel, a complex of structures built during the course of several centuries and encompassing a broad range of architectural styles. The abbey visually towers over the island, the village, and the walls situated at its lower elevations. Whereas once visitors to the island had to reach it by boat (if the tides were in), or by foot across the sands and mudflats, a causeway built in the late nineteenth century now permits easy access. Mont Saint Michel is famous, among other reasons, both for the rate at which its tides flow in and the depths achieved by them at certain times of the month. At the height of the tourist season, the island's single street is clogged with visitors who traverse the site on foot.

What is it about Mont Saint Michel that draws people to it? What makes this space so meaningful to them? There are many reasons, perhaps as many as the visitors who travel to Mont Saint Michel every year. The fact that it was designated a World Heritage Site by the United Nations Educational, Scientific, and Cultural Organization (UNESCO) in 1979 undoubtedly adds to the number of reasons Mont Saint Michel is such a desirable place to visit. Yet it is the abbey, to which Christian pilgrims have traveled for centuries, that dominates the island. The religious significance of Mont Saint Michel—named in honor of the Archangel Michael, who is represented as a gilded figure atop the abbey's spire—cannot be denied or ignored. The meaningfulness of this space, thus, is related to and even depends on historical and religious significations and to the ways in which religion (in this case Christianity) imagines, creates, shapes, and represents its space(s).

The importance of religion in the space of Mont Saint Michel is inescapable. The abbey is visible from some distance away and is built on a grand scale. It is a massive structure of walls, abutments, windows, roofs, towers, and spires. Farther down the island's slopes are the buildings of the village, all of which are dwarfed by the abbey. At the island's base is a stone wall, built to defend the island from invaders. Two gates provide access to the village and the street that leads visitors to the abbey. Hotels, restaurants, and shops line the street, catering to the material needs of the people in the island's space(s), thereby acknowledging that not all needs are religious or spiritual. The climb to the abbey is steep, but it offers an excellent view of the surrounding area to those who attempt it. Depending on the time of day, visitors can watch the incoming tide that transforms Mont Saint Michel once again into an island.

Mont Saint Michel stands at the convergence of different natural and social forces. The famous tides are influenced by the gravitational pull of the moon and sun on the earth. Those tides also helped create both the island and tidal flats by washing away materials over time to expose the rock and create the flats. Storms, climate, and other natural factors likewise contributed to shaping the island. Further changes were made by the island's inhabitants, who created spaces suitable to their purposes and needs. The designers and builders of these spaces had to account for topographical realities, such as the island's slope, elevation, terrain, and other physical properties when building or creating spaces of meaning and significance.

In addition to these natural and social forces, Mont Saint Michel is imbued with a number of religious meanings and representations. The architecture, for example, uses design, materials, orientation, high ceilings, distribution of light, and other built features to represent and embody ideas about sacred space and the link between transcendent and immanent worlds. The location of the abbey and other buildings at the top of the island, symbolizing proximity to heaven, is an expression of Christian ideas and understandings about the nature of the cosmos, as it enables worshippers to be closer to God. So too was the placement of the transept of the church at the top of the island and the various other buildings of the abbey around it.

The origins narrative of Mont Saint Michel is another example of its religious significance. It is said that in 708 C.E., the Archangel Michael commanded Aubert, bishop of Avranches, to build a church on the island. The recalcitrant Aubert complied only after Michael burned a hole in the bishop's skull. Mont Saint Michel is a destination for Christian pilgrims who travel (and traveled) there for reasons of religious devotion, piety, and practice, thereby adding another meaning to its spaces. The closure of the abbey after the French Revolution and its conversion into a prison added further meaning to the island's spaces, transforming them from sacred religious spaces to civil spaces. In turn, the closure of the prison in the mid-nineteenth century, because Mont Saint Michel was deemed a historical architectural treasure, gave new symbolic meaning to the island and its spaces, this time involving national heritage. The history of the site thus comes to play a role in its social meanings.

Mont Saint Michel receives over three million visitors each year and is one of the most visited Christian sites in Europe, facts that add to its meaning, as (at least) a popular tourist attraction. The symbolic meanings of Mont Saint Michel in the twenty-first century are not limited to Christianity; tradition and its religious representations and ideas also play a significant role in them.

Reading Meaningful Spaces

The spaces of Mont Saint Michel encapsulate many of the themes and issues addressed in this book. The physical spaces of the island are constituted by both its built structures and its natural environment, ranging from the rocks and slopes of the island to the tides, weather, seasons, and erosion. These realities link the island's spaces not simply with one another, but with the larger universe, as gravitational pull affects the tides and water surrounding it. The causeway (soon to be replaced by a new causeway and light bridge[1]) makes access to the island easier these days (although many tourists walk on the mudflats and tidal plain), yet the island's walls, gates, and single street up to the abbey create an awareness of how the body moves through its spaces. So too do the different architectural styles of the abbey and the supporting buildings (Romanesque and Gothic), to say nothing of the contrast between those styles and the ones used for the shops, restaurants, and other structures and spaces of the town. Such styles change over time as building technology, tastes, and architectural thinking advance. These developments also reflect and affect understandings of the nature of the town and its relationship to the abbey (certainly the UNESCO World Heritage designation affected these ideas; see Jacob Kinnard's analysis of how Bodhgaya similarly was affected). These conceptions of architecture, the town, the abbey, how to improve access to the island, and other matters all involve mental space, in the sense of being ideas about physics, materials, space needs, planning, and similar concerns. Yet concepts of space also involve the ways in which people actually live in space. Few people live with blueprints or other schematics of space constantly in mind, but rather they come to live the spaces they use in terms of the symbols, myths, and social meanings they associate with space. That Mont Saint Michel is a Christian pilgrimage site, for example, moves beyond the plan for the island to significations influenced by Christian theology and practice. The abbey was built at the top of the island as part of an expression of Christian ideas and practices, which understood heaven to be above the earth and sky.

There are a number of ways in which social spaces such as Mont Saint Michel may be analyzed so as to understand better religious representation in this place and how religion and space intersect to create and produce meaning. As the chapters in this book make clear, many possible meanings are generated by the interaction of space and religion. These chapters take up the question of meaningful space from a range of disciplinary locations, and thus with a variety of different assumptions, methods, concerns, and interests. They also utilize a broad range of understandings of religion, be that in terms of an existential orientation, a meaning-making communication system, rituals and practices, faith and devotion, or theology and scholarly discourse. Notwithstanding these disciplinary differences, however, they have numerous points of intersection. This is due in no small measure to the fact they originated in an international conference held in Zürich, Switzerland, in 2011. The conference, "Meaningful Spaces," was interdisciplinary in participation and brought together scholars from academic fields as diverse as developmental pediatrics and Hebrew Bible, urban planning and film studies, astrophysics and religious studies. What holds the variety of contributions in this book together is their focus on meaningful spaces and a series of questions about this topic that were addressed during the conference.

Of particular concern is the question of how religion serves as a system of representations generating space and associating meanings with it, in recognition of the fact that space is a fundamental aspect of human existence, perception, behavior, and action. The chapters consider this question in terms of the conceptions and representations of space within religious symbol systems and how those systems influence physical space, spatial understandings, and the meanings spaces are given. Space thus serves as a lens through which to analyze religion and religious systems, just as religion and religious systems serve as a lens for analyzing space.

The chapters in this book are organized in three sections, beginning with conceptual and theoretical issues before moving to case studies, which are grouped into those from the ancient and medieval periods and those from contemporary periods. Section I articulates a set of theoretical approaches that play out in the rest of the volume. These include questions of how human beings come to have a concept of space, of the ways in which spatial concepts are understood and articulated in human terms, how space is to be seen, named, and depicted, and what meaning space has for humanity, if anything. As is to be expected, these chapters demonstrate that religion has a significant influence on these questions and, conversely, space and conceptions of space influence religion and religious ideas.

Section I begins with Remo H. Largo's chapter, in which he examines the development of spatial cognition and awareness in young children. A developmental pediatrician, Largo details the results of the research he undertook to determine the ages and points of development at which children become aware of spatial distinctions (such as between a child and its mother), tracing that progress to the point at which three-dimensional understanding is achieved. His chapter is followed by that of Albert O. Benz, an astrophysicist who takes up the question of whether, and how, the physical universe has any meaning. Over time, as astronomers gained better and more precise definitions of the universe, its size, scope, and functioning, understandings of the size of the universe grew exponentially. In the process, the question arose as to whether space has any meaning for humanity, and this is a question that has persisted for four centuries. Because the question of meaning involves interpretation, it is not particularly well suited to physics and scientific explanation. The epistemological and interpretive understandings of religion, on the other hand, do offer answers. Metaphors of creation and the idea of *creation continua* offer interpretations of the universe though not explanations of it. Benz shares with Largo a concern for examining how humans, from the ancient Greeks to modern scientists, gain an understanding of space and conceive of the universe, the world around them, and their place(s) within it.

One of the ways in which humans understand and represent their place in the world is through maps, which are the focus of Franco Farinelli's chapter. Tracing their development through history, he explains how Christian theology and ideas influenced early mapmaking and maps in the Western world, such as locating Paradise on them. Even with the emergence of geography and the changing conception of the world it signaled in the modern period, Christian theology and ideas continued—and continue—to exert influence over maps. Farinelli answers the question of meaning in space (Benz) by noting how Christian ideas were fused with the very essence of space.

Carla Danani's chapter concludes this section by bringing a philosophical focus and analysis to the question of human dwelling, place, and space. Observing the importance for human beings of a place of belonging, of existing "somewhere," Danani makes an argument for the significance of spatiality for a philosophy of experience, ethics, and politics. She does so by asking not only how space is treated in an ethical way, but how ethics and politics are shaped by space. Her chapter brings to the fore the importance of defining what "space" is being considered, which "place" is assumed, and who is speaking, acting, and analyzing (a concern also present in Farinelli's chapter). Such questions are an aspect and ability of human rationality (cf. Largo). She explores the human condition as being a located and symbolic reality before considering the implications of such an understanding for ethics and politics, where the "here" of human existence permits a correlation between subjective and objective space. Danani's chapter resonates with aspects of Benz's work, since her analysis of the importance of space for ethics, politics, and behavior poses a question of orienting oneself in the world. This is a question Benz also addresses, albeit in a different way and scale. Yet because they ask the same basic question, it links these two arguments. So too does her investigation of how human beings derive an understanding of subjective space from objective space. Both Danani and Benz are concerned with how human beings express the meaningfulness of space, a subjective experience.

Sections II and III turn from the larger conceptual and theoretical issues of Section I to focused examples of the intersection of religion and space. Informed by theories of space articulated in fields including philosophy and sociology, these case studies analyze examples spanning social, cultural, and historical contexts, as well as diverse media. These cases represent space and religion to the viewer or reader in particular ways, be that in the construction of the south portal of the Chartres cathedral or through artistic images. The authors analyze how space, religion, social issues, history, and the media employed, among other factors, intersect in some meaningful way in the materials under consideration. As these factors are identified and examined, questions and issues from Section I reappear, including how space is conceived as it is being represented, how cosmological ideas inform space, how space has meaning for humanity, and the social, political, and ethical implications of space for society and individuals.

Section II contains chapters focusing on ancient and medieval case studies. Mark K. George examines the description of ancient Israel's wilderness tabernacle in the Old Testament/Hebrew Bible and the connection it has with creation and the cosmos. George considers the ways in which the human body is assumed in tabernacle space to argue that its connection with that space ascribes to the body cosmic meaning and significance. The tabernacle symbolically is a microcosm (micro-cosmos) of creation, and the body's connection to it conveys that symbolism. Since the tabernacle is capable of being repeatedly oriented within the cosmos, so too are the bodies of the Israelites. In this way the cosmos becomes a meaningful space, and individuals understand their places within it—themes and issues already expressed by the chapters in Section I, such as how one orients oneself in space (Danani), how there is meaning in the cosmos (Benz), and the connections between maps and theology (Farinelli). These themes and issues are engaged in a different way in Christl M. Maier's chapter, in which she uses the spatial theories of Henri Lefebvre in a close reading of the eighth century BCE prophet Hosea. Maier argues that Hosea's polemical representations of local sanctuaries and worshippers in gendered terms is part of his attempt to distinguish Israel's deity from other deities of the region, arising out of a confusion on the part of his audience as to which deity is providing for the audience's welfare. She analyzes Hosea's spatial practices, conceptions of space, and the meanings he ascribes to spaces to highlight the spatial nature of his polemic against his opponents. Attention to spatial aspects of human experience and how they are represented in the biblical narratives enables her to explain how the prophet puts forth an alternative spatial understanding and explanation for his audience. Maier's chapter engages the question of space and the human body that George and Danani previously discuss, expanding that theme in new directions to include gender. She anticipates the chapter by Katherine Rousseau, the next in this section, by examining how the body interacts with particular spaces.

Rousseau's chapter signals a change in history, social context, and media. The function of apocalyptic space in the Last Judgment portrayed in the central portal tympanum of Chartres cathedral's south transept, and its effects on medieval spectators and pilgrims to this site, is the focus of her case study. Rousseau considers how imagined spaces of the apocalypse come to be represented in art and life and the effects such representations have on individuals. She examines the depiction of the Last Judgment in the portal and the textual, spatial, and theological influences evident in it, before turning to consider how that space was lived in the thirteenth century CE. Her analyses permit her to argue for the ways in which this representation of space and religion had rhetorical, social, and political significance with respect to both the community of Chartres and ideas of the cosmos. Rousseau's analysis takes up many of the themes and issues found in the other chapters, including cosmology (Benz, George, Pezzoli-Olgiati), the presence of human beings in space and interacting with spatial and religious representations (George, Pezzoli-Olgiati, Kinnard, Mendel, Fingerhuth), the impact of space on ethics and politics (Danani), and even how the representation of apocalyptic space serves as a map of reality (Farinelli). Daria Pezzoli-Olgiati's interest in afterlives and "other worlds" explores a similar issue, namely how such ideas can be represented when they are beyond human experience and thus inaccessible to the living. She examines two artistic representations of other worlds, from the fifteenth to seventeenth centuries in German-speaking regions, in which worlds beyond the boundaries of human life are visually represented and analyzed. Pezzoli-Olgiati's interpretation of these examples considers not only the images themselves but how the viewer creates a particular type of space and experience of it. Her chapter indirectly provides an answer to Benz's question of how the inexpressible can be expressed, although in her examples, it is not the scale and meaning of the cosmos that must be expressed, but worlds beyond the cosmos (see also the chapter by Farinelli). Pezzoli-Olgiati also is attentive to the reality that human bodies are both in space and create space and thus adds to the larger discussion in this book (Benz, Danani, Farinelli, George, Rousseau).

Section III begins with S. Brent Plate's consideration of the ways in which ornamentation and ornamental imagery are used and understood, then reused and understood in a different way, by succeeding groups who occupy the same spaces. He examines a variety of structures in Andalusia, Spain, and the ways in which ornamentation from an earlier historical, social, and religious period is reused in later historical, social, and religious periods. This historical review enables him to inquire about the ways in which space is produced by groups and societies, as the earlier ornamentation on these structures is reinterpreted and refigured to the needs of those who follow them. Such reinterpretations, however, never totally erase earlier significations and meanings of those ornamentations. This chapter is another articulation of the theme of the relationship between the viewer and how space is represented and signified (Rousseau, Pezzoli-Olgiati, Glavac, Mäder).

Monika Glavac examines eighteenth-century English caricatures, which were quite popular in English society at the time. Her examination leads her to consider larger geographic, social, and economic issues related to the production, display, and consumption of caricatures in English society. In this way Glavac is able to consider both how space is represented within caricatures as well as how these caricatures created space for their viewers and consumers. A careful analysis of one of William Hogarth's caricatures of Methodists and Methodism explains how Hogarth's commentary operates and how he positions the viewer in relation to it. Glavac's analysis shares with other chapters in the book an interest in the role and position of the human body in space and how the body creates space, the politics of space, and the relationship between space as represented in media and the one viewing that media. These are concerns shared by Marie-Therese Mäder, who considers space in film and with viewers of these films. The creation of such a space plays an important role in understanding how meaning is created in the experience of film. While there are multiple ways filmmakers and their films attempt to create that space, Mäder focuses on filmic journeys, using two films, *Bab 'Aziz* and *Mar Adentro*, as examples of this process. She analyzes how these films involve viewers in creating meaningful space. Mäder shares with Glavac, Pezzoli-Olgiati, Rousseau, George, and Danani a consideration of the body in space and the ways in which spaces become meaningful to individuals. This is an issue also addressed in the chapter by Anna-Katharina Höpflinger, in which she considers how clothing both creates and marks particular spaces. These spaces are not socially or economically neutral; as Höpflinger argues, spaces created by the clothed body are clearly demarcated in social, economic, and geographic terms. Clothing also creates and marks religious space, as demonstrated by Höpflinger's examination of the Scaletta cloak. The Scaletta cloak is worn by Protestant clergy in the Swiss canton of Grisons during religious occasions and is particular to them. Wearing this cloak creates meaningful space for the wearer as well as those participating in the religious occasion. Höpflinger recalls the clothing worn in Israel's tabernacle when the priests entered tabernacle space, or when the high priest entered the most holy space (George), and suggests another way in which bodies create particular meaningful spaces.

Journeys in space are at the heart of Tommi Mendel's chapter, which focuses on modern-day pilgrims and the stages pilgrims go through on their journeys. Pilgrimage involves more than simply travel on the road to and from one's destination. As Mendel argues, it involves a series of stages that include preparations for the journey, the journey itself, and reintegrating into one's post-pilgrimage life. Modern pilgrimages also involve multiple spaces, in no small part due to the proliferation of travel and guide books that recount for pilgrims (potential and actual) the places they will visit, the sights they will see, and the experiences they will have. Central to Mendel's argument is that (pre-) conceptions of spaces play a significant role in pilgrimages. They shape the expectations and perceptions of pilgrims before, during, and after their travels. Therefore, what makes the pilgrimage meaningful is that the sites and sights the pilgrim encounters conform to those of other travelers, not that they have their own particular, unusual, experiences. Mendel's chapter complements those of Rousseau and Pezzoli-Olgiati, in the sense that each examines the encounter of an individual with a space, object, or representation that seeks a particular response from the viewer. The Last Judgment is a warning to those passing through the south portal; visual representations of other worlds and afterlives seek to warn viewers about fates beyond life, and pilgrims tend to conform their perceptions to the descriptions in travel books they read in preparation for their journey. The journey theme also complements Mäder's chapter, as a counterpoint to how such journeys might be experienced. Finally, Mendel's chapter complements Kinnard's arguments about UNESCO sites, as both address questions about perceptions of space, the idealizations of space, and the ways in which the reality that spaces change over time gets ignored by certain users of these spaces.

Carl Fingerhuth raises questions about how the character or *Gestalt* of a town or region might be expressed in architecture in the period he terms "beyond the modern." Drawing on his life's work and experiences, his chapter is a meditation of sorts on the practice, conception, and symbolism of spaces. Using the past to look both to the present and the future, Fingerhuth sees humanity moving into the fourth and, eventually, the fifth dimensions as it evolves and he argues its spaces too must reflect that development. Just as spaces in Andalusia combine elements of the past with the present according to Plate, so too does Fingerhuth argue that all cities and towns engage in this practice, which presents a host of challenges for architects and urban planners seeking to (re-) develop city spaces in ways that capture the time within which that development occurs. Both the past of a city or town and its present and future must be expressed if such spaces are to be, and remain, meaningful to their users and inhabitants. Those meanings must capture and articulate the spirit of cities and towns and connect to the new energies within them. Due to its range and scope, this chapter touches on a number of themes and issues expressed in the other chapters in the volume, including how space is understood (Largo, Benz), the role of the body in producing and experiencing space (George, Maier, Rousseau), the connection between religion and space (Farinelli), the role of the viewer in space (Pezzoli-Olgiati, Mäder), the use of the past in creation of the present (Kinnard), the connection or harnessing of new energies and symbolisms in space (Plate), and the effects of space on social, political, and ethical issues (Danani).

The chapter by Jacob Kinnard completes this section. In it he examines attempts by UNESCO to preserve specific places and spaces considered to be of great significance for humanity. Specifically, he considers the case of Bodhgayā, the site in India recognized by Buddhists as the place where the Buddha attained enlightenment. The temple at Bodhgayā is, as Kinnard notes, a living temple, a place and space that has experienced ongoing change over the past 1,500 years as Buddhists and Hindus occupied, abandoned, built, demolished, rebuilt, and changed the site. The 2002 designation of the site as a World Heritage Site (WHS) by UNESCO was celebrated by Buddhists even as it juxtaposed modern, Western ideas of space over those of the people and religions actually using Bodhgayā as a temple. Kinnard considers the implications of this juxtaposition for Bodhgayā and those who use it in their religious practices. The ideas expressed by him serve as a complement to Plate's arguments, as both take up issues of living religious spaces. Kinnard complicates Fingerhuth's consideration of the *Gestalt* of a town or region, pointing out how a site, town, or region may be changed or influenced in unanticipated ways by outside influences.

Pierre Bühler's chapter concludes the book. In it, he provides a review and reflection on all the chapters from a hermeneutical perspective. Taken as a whole, the chapters in the book create a space of interpretation as a result of being interdisciplinary and interacting with one another. Bühler draws upon the work of Blaise Pascal, Martin Heidegger, Maurice Merleau-Ponty, and Paul Ricoeur to highlight the hermeneutical themes in the book, including the fundamental spatial reality of human existence, that human life is lived in a dynamic relationship with space, and of the need to make sense of human existence within space despite being incapable of gaining sufficient distance from it adequately to understand it. Religion and its hermeneutical practices (particularly, for Bühler, within Christianity) are one means by which humans represent space, live in it, and come to understand it. Yet whether interpretations of space are philosophical or religious, they are rife with tensions. The chapters in this book do not resolve these tensions but participate in them and propel them forward.

Meaningful Space as an Interdisciplinary Challenge

The question of what makes space meaningful—be that Mont Saint Michel, the universe humans inhabit, or the act of viewing a film or painting—is a rather complicated question to answer. The essays that form the chapters in this volume are a beginning, but certainly not the end, since there are many more ways in which the question might be answered. The work of Largo on developmental aspects, for example, may be complemented by the recent and emerging work in religion, biology, and neuroscience, in which researchers seek to understand what happens physiologically as people are "religious" in some way. Economic and psychological answers are possible. And the list goes on. We are convinced there is a great deal of potential in pursuing additional interdisciplinary work on the intersection of religion and space, because the chapters in this book suggest that such work is beneficial. Defining such terms as "religion," "space," and "place" must be part of interdisciplinary conversations. Nevertheless the benefits of bringing together scholars from such a range of disciplines proves to be informative and enriches the conversation by providing new ways of answering the question.

Note

1. "Access Works," La Baie website, *Syndicat Mixte Baie du Mont-Saint-Michel*, accessed June 29, 2013, http://www.projetmontsaintmichel.fr/en/works/access_works.html.

Section I

Theoretical Frameworks for Approaching Space

Chapter 1

How Children Develop an Understanding of Space: A Developmental Pediatric Perspective

Remo H. Largo

A child experiences space in the first few years of its life in a great variety of ways. During the mother's pregnancy, it moves in the womb under weightless conditions. After birth it has to fight against the force of gravity. In the course of the first year of life, it learns to grasp objects and to move in space. In the second year, it begins to understand spatial relationships while playing with objects. Finally, it links its conception of space with linguistic terms and an understanding of language for spatial interrelations starts to grow.

Sitting Up and Moving

The unborn child begins to move in the eighth week of its mother's pregnancy. The first experiences of space are very special; the child is almost weightless in the amniotic fluid and therefore able to move in all three dimensions of space. It can turn on its own axis, walk in the amniotic sac, and even do somersaults. There are similarities between the movements of an unborn child as we observe it via ultrasound and that of an astronaut walking in outer space.

At birth a child is abruptly placed in a very helpless situation due to the force of gravity. It kicks with its feet and waves its arms about, but can scarcely lift its head. Sitting up straight or moving forward is impossible for months. If it wants to change the position of its body, it is reliant on the help of its parents.

The first gross motor skill an infant acquires is controlling the position of its head. An infant can raise its head against the force of gravity and hold it straight when lying on its stomach and afterwards when sitting up too. Between six and twelve months, posture control extends to the whole body. At about nine months it can sit up unaided. This opens up new possibilities for the child. Its perception of the room is much more subtly differentiated. It can participate in what is happening in its near surroundings better than when lying down.

Between five and seven months of age, the day comes when a child changes its position for the first time. It turns itself from lying on its stomach to on its back and, a little later, from on its back to on its stomach. At seven to ten months the child begins to crawl forward on its stomach. In the beginning it uses only its arms. It supports itself with its hands or elbows and pulls its body forwards. After a short while it begins to use its legs as well. Between 10 and 18 months most children take their first steps. They not only want to walk but are also eager to explore their environment and experience space. From week to week they are able to reach more and more places within their immediate environment.

Once the child has its motor activity sufficiently under control, it begins to get interested in all possible kinds of toys and means of locomotion. Slides, swings, and towers for climbing magically attract three- and four-year-old children. Various moving objects, like a small tractor or a tortoise on which the child can sit and push itself forward with its feet, are similarly popular. Children learn to ride a tricycle at about two-and-a-half to three years of age. A child on a tricycle makes us very aware of how skillfully it can move in space. The legs deliver the power to move forwards, the arms steer the tricycle and the whole body keeps in balance, so that the child does not fall off. Thus, the child has its entire motor activity sufficiently under control to adjust speed and changes in direction according to spatial conditions. For this, it requires a differentiated conception of space and a distinct sense of the best way to move within it.

Playing with Space

An infant also experiences space when playing with its own hands. If we watch a newborn baby awake as it plays with its hands, we can observe how it occasionally brings its hand in front of its face, opens its fingers, moves them slowly, and constantly examines them. Hand-eye coordination becomes well developed in the second and third months of life. The child now has its arm, hand, and finger movements under control to such an extent that it can deliberately touch an object. When a child begins to grasp objects at four or five months

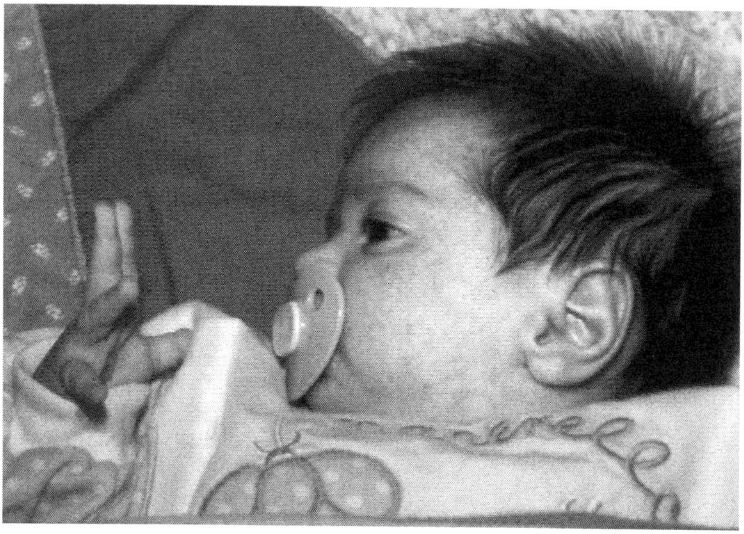

Figure 1.1 Hand-eye coordination. A two-month-old infant looks at its tiny hands.
Source: Remo H. Largo, Childrens Hospital, Zurich.

of age, it localizes the object in space, guides its hands toward it, and finally grasps the object (see figure 1.1).

From then on, the child plays intensively with objects. Some forms of play between two and six years of age reflect the child's rapidly growing understanding of space. They give us an insight into how the child deals with the spatial relationships of objects and the dimensions of space and gravity. Play is characteristic of the corresponding development age of a child. Thus, at the age of 12–16 months a child is greatly interested in containers and their contents. At 16–20 months it likes to stack objects on top of each other. These various forms of play appear in all children in the order described below (Largo et al. 1979a).

Container Play

In our studies on play behavior, we used a tiny glass bottle, which contained small wooden balls. Using this bottle, the development of container play can be followed. A six-month-old child immediately takes the bottle to its mouth. It does not notice the little balls, even when made aware of them by shaking the bottle. The child perceives the bottle only as a whole. After seven months of age, the child shows a growing interest in the balls. It tries to reach them with its finger through the glass. At nine to twelve months of age, it sticks its forefinger in the opening of the tiny bottle and tries to get at the balls. The child has developed an understanding of how one object can be inside another object (see figure 1.2 a–d).

At 12 months of age, the child puts the balls in the bottle but is unable to take them out again. If we demonstrate how the balls can be taken out of the bottle by tipping it, the child is not yet able to imitate the (tipping) action. Most children try to get the balls out by shaking the bottle. After being shown how, a child can tip the balls out of the bottle at about 15 months. At 18 months it empties the bottle spontaneously.

Between 10 and 18 months children show a great liking for filling and emptying containers of all kinds. The child's interests are not focused on playing with beakers or cubes, but rather on cabinets, kitchen drawers, CD collections, and book shelves—to the annoyance of its parents. Parents can make life a little easier for themselves if they give the child two drawers which it can fill and empty.

Figure 1.2 (a–d) Tiny bottle with small balls. (a) Eight-month-old Tabea notices the balls in the bottle and tries to get them through the glass. (b) Twelve-month-old Leandro knows that a container has an opening. He tries to reach the balls with his finger. (c and d) Eighteen-month-old Rhea empties the bottle by tipping it and puts the balls back with her hands.

Source: Remo H. Largo, Childrens Hospital, Zurich.

Figure 1.3 Vertical building.

Source: Remo H. Largo, Childrens Hospital, Zurich.

Vertical Building

Between 15 and 18 months a child begins to pile up objects (see figure 1.3). It not only builds towers with cubes and building bricks, but also stacks up things that are less suitable for it, for example, the doll house chairs.

Horizontal Building

At 24 months, interest in building towers begins to wane. The child enthusiastically makes rows of blocks. It now shows a preference for horizontal building. The child is fascinated by toy train sets, for example, where the carriages can be linked together.

Combining Vertical and Horizontal Building

At about two-and-a-half years of age, a child brings the vertical and horizontal dimensions of space together in its play. For example, it builds stairs for a doll, or a bridge under which a train can run, with building blocks (see figure 1.4 a–b).

Three-Dimensional Building

At the age of three to four years the child begins to build rooms, in that it takes into consideration all three dimensions. For instance, a young boy might construct a garage for his cars with building blocks. By the age of five, the understanding of space has developed to such a degree that children are able to build houses, airplanes, and automobiles with Lego or other materials.

Figure 1.4 (a–b) Spatial play. Bridge at two-and-a-half to three years (a), a house at five to six years of age (b).

Source: Remo H. Largo, Childrens Hospital, Zurich.

Developing a Schema of the Body

An understanding of space is not only important in gross motor development and play, but also in the perception of one's own body. At 18–24 months, a child begins to identify some parts of its body, such as the nose, hands, or legs. At three or four years of age, it develops a comprehensive idea of the human body, which it also applies to other people and to dolls. For instance, if a child wants to dress a doll, to pull a dress over its head and put on shoes, it uses its own schema of the body.

Drawing

Prior to four years, a child lacks the fine motor skills necessary for translating its mental images of objects and living things into a drawing. At three years of age, it may already have a differentiated idea of the human body, but it cannot yet draw the human form. It can only imitate the movements that older children and adults make when writing and drawing with a pencil. By four years of age, the lineal scribbling turns more and more into round, closed forms. Then the child manages to draw the first picture of a human being, this is the so-called pre-symbolism phase. By the age of five, the child can extend this round form into head, neck, trunk, and extremities and add many more details, like hair, hands, and feet. At kindergarten age, the drawings of people contain all important elements of the human figure, even if the proper proportions are still missing. A child depicts the human form almost without exception from the front and only rarely from the side view. If it draws several forms, for instance its family, then the size and details of the individual figures are often less related to their actual size than to the emotional significance which

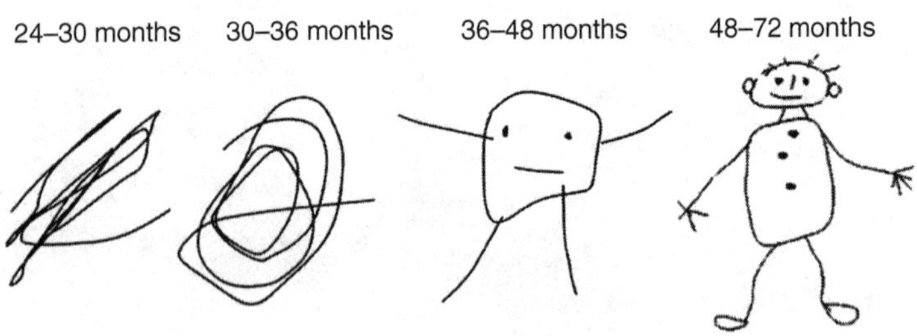

Figure 1.5 From scribbling to drawing the human figure.
Source: Remo H. Largo, Childrens Hospital, Zurich.

these have for the child. Other drawings, such as those of houses, automobiles, or animals, are also done, but they remain rare up to five years of age (see figure 1.5).

Naming the Space

Children at all ages understand cognitively far more than they are able to express linguistically. Thus, the understanding of space, as it is expressed in play, is always clearly ahead of linguistic understanding (Largo et al. 1979b, see figure 1.6).

First of all, the child comes to know the meaning of the preposition "in." Already at the end of the first year of life, it learns that an object can be inside another object. In so-called container play, the child examines this conception of space. At the beginning of the second year of its life, it understands that the preposition "in" stands for a distinct spatial relationship between two objects. If its mother says: "The apple is in my bag," the child searches for the apple in its mother's handbag. Weeks and months pass before the child can also say the word "in" and use it. Most children do not do so before the end of their second year.

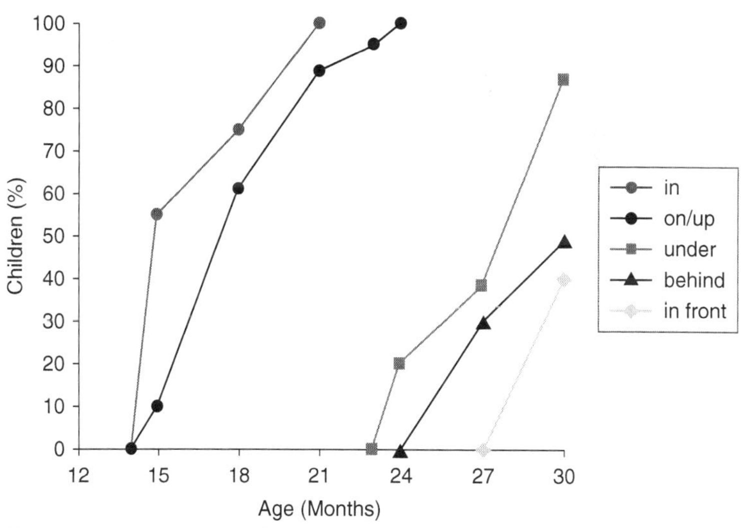

Figure 1.6 Development of the meaning of the preposition "in." The curves represent the number of children (as a percentage) who play with containers at a certain age, and who understand and finally use the preposition "in."

Source: Remo H. Largo, Childrens Hospital, Zurich.

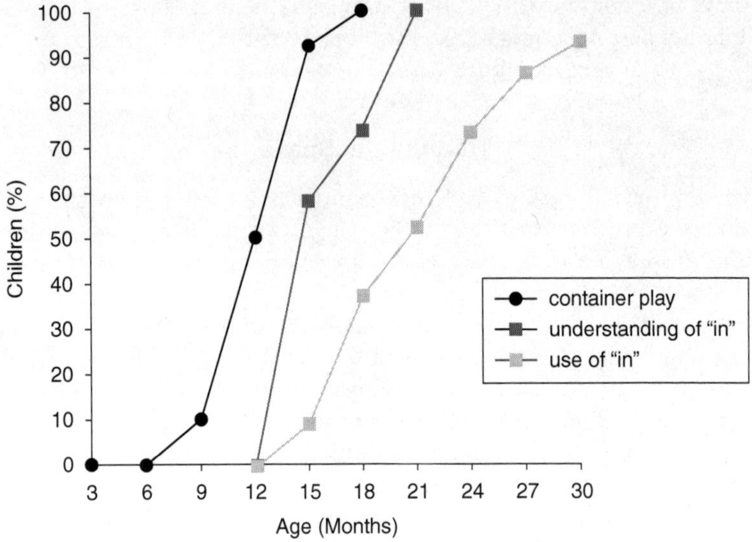

Figure 1.7 Development of an understanding for spatial prepositions.
Source: Remo H. Largo, Childrens Hospital, Zurich.

In the following, the order in which the spatial prepositions occur is given. This sequence is essentially the same in all children. After the word "in" follows the word "up." If they want their mother to pick them up, they say "up"—but they often say the same when they want to go back on the floor. With "up" they mean a vertical movement. After the preposition "up" follows "down," then at two-and-a-half to three years of age "behind" and finally "in front" (see figure 1.7).

When a child reaches school age, development of its understanding of space is still not complete by any means. For instance, it still has to learn to understand that the earth is a sphere, to grasp its size in numbers, and to get an idea of the immensity of the universe and the billions of stars. A basic understanding of spatial relations, however, is already acquired by the child in the first few years of its life.

Chapter 2

Meaningless Space?: Astrophysics and Theology

Arnold O. Benz

The size of the universe has always been beyond human imagination. Our scales, be it the human body or the distance of a day trip, are many orders of magnitude smaller and useless in comparison with cosmic dimensions. This inhibits relating human space and cosmic space. How can we orient ourselves in this unimaginable cosmic space? The lack of relation and orientation leaves us without firm ground under the feet; the question of meaning is lurking within fathomless space. This condition is not new. The threat of meaningless space has been with us now for four centuries. In this chapter I demonstrate that the universe's ten billion light-year size is a necessary condition for our existence. Does this make space meaningful? Meaning, I argue, does not result from scientific explanations, but from an interpretation in a larger frame of reference. A theology of space interprets space as a gift of creation, establishing a relation between Creator and creatures.

Frightening Space

In disputes between the geocentric and heliocentric world models in the sixteenth and seventeenth centuries, the meaning of space was a major issue. Cosmic space was spanned by planetary orbits and the respective motions of the Sun or Earth. In the geocentric model of Claudius Ptolemy, Earth was at rest, but the Sun moved in a circle. Nicolaus Copernicus proposed in 1543 a model where the Sun was at the center and the Earth moved around it like the other planets. In both models, the stars were fixed on a sphere just beyond Saturn at a distance estimated at the time at ten to twenty thousand Earth radii. The astronomers of the seventeenth century underestimated the distance to Saturn by a factor of a hundred.[1] Their model cosmos was tiny compared to the present conception. Its limit at the stellar sphere was at only one hundred thousand days' journey by horse. It would fit easily within a tenth of the orbit of Mercury, the innermost planet, as we know it today.

Tycho Brahe (1546–1601) and other astronomers of his time rejected the heliocentric model proposed by Copernicus. At that time, parallax motions of stars due to the yearly orbital motion of Earth were not observable. If stars were at the distance of Saturn or just beyond, their apparent position in the sky would draw an observable circle in the course of a year while Earth was making an orbit. Brahe, the best observer at the time, did not observe such a motion. Thus, he concluded correctly, that the Earth did not move around the Sun, or the stars must be at least 700 times farther away than Saturn. The heliocentric model would require a gigantic space beyond Saturn. Such a size was not conceivable for Brahe. It would be much larger than required for the stage of the human drama, believed to be the purpose of the universe. Thus, the empty space beyond Saturn would have *no meaning*. How could God have created such an enormous universe, when only a minute part of it would be useful? So Brahe developed a model with a stationary Earth in the center encircled by the Sun, while the planets orbit the Sun. This compromise left the stellar sphere not far beyond Saturn and was consistent with observations until the late seventeenth century.

At the time of Blaise Pascal (1623–1662) the heliocentric model was getting more and more established among scientists. However, he experienced the progress of science not as a triumph of human cleverness, but as an indication of his insignificance. He notes: "The eternal silence of these infinite spaces frightens me."[2] The silence may be understood as the absence of any relation of this additional space with mankind. He could not imagine how the enormous space between Saturn and the stars had any meaning. We note here that meaning is not a scientific term based on measurement and causality, but expresses a relation. Pascal perceived the absence of any relation to humanity and thus without meaning as frightening. If most of space were without meaning, the meaning of the whole cosmos and even of humanity would be questionable.

When Friedrich Wilhelm Bessel measured the first parallax in 1838, he discovered that the nearest stars are not 700 times but about a million times farther away than Saturn. The distance to the closest star, Proxima Centauri, is measured today at a distance of six billion ($6{\cdot}10^9$) Earth radii. If Earth were as large as a pinhead, the nearest star would be the size of a cherry at a distance of 4,000 km. In today's view of the world, Brahe's problem is actually more than a factor of a thousand times worse than he feared.

The size of the accessible universe made another jump in the 1920s and increased by an additional factor of a million when the distance to the Andromeda galaxy became measureable. Finally, a further factor of ten thousand came into play with the discovery of the microwave background in 1965. It originated at the time when the early universe became transparent, 370,000 years after the Big Bang. Since that time radiation was not absorbed anymore and it propagated freely in space. As the universe was opaque before, this surface 13.7 billion ($13.7{\cdot}10^9$) years ago is today's horizon for visual light. Since the time when background radiation was emitted, the surface has moved to 45.7 billion light-years from us and constitutes the limit of the observable universe today. It is at a distance of a hundred million billion (10^{17}) times the alleged radius of the stellar sphere (and thus of the universe) at the time of Tycho Brahe. Cosmic space has grown from the still unimaginable distance to Saturn by an unimaginable factor.

Space in Today's Universe

The size of the observable universe is related to its expansion with nearly the speed of light since the Big Bang 13.7 billion years ago. Space began to expand rapidly beginning with a phase known as "inflation" and continues to be accelerated today, driven by enigmatic Dark Energy. This energy comprises three quarters of the cosmic energy; however, this concept is very poorly understood. It may be related to the energy that has initially driven the Big Bang. There are several controversial theories on the nature of Dark Energy. One of them relates it to quantum fluctuations in a vacuum yielding a nonzero energy density. In any case, without this energy, space would not expand and the universe would still be hot and dense. Thus Dark Energy is essential for the further development of the universe into galaxies, stars, and planets. Space is not empty, but overflows with Dark Energy everywhere. Dark Energy in space has a relation to us humans because it helped to shape cosmic space where galaxies, stars, planets, and finally life could develop.

Galaxies form under the gravity of Dark Matter, another enigma of cosmic space not related to Dark Energy mentioned earlier. Dark Matter is an unknown form of matter that can be observed by its effect of gravity. Thus Dark Matter has mass, but negligible interaction with ordinary (baryonic) matter and therefore it cannot be studied from collisions. Particles yet unknown are suspected to comprise this Dark Matter. These particles dash through space without taking notice of normal matter and are not influenced by anything other than by gravity. Thus they do not emit light or damage molecules in living creatures. Millions of Dark Matter particles traverse our body per second. Space is full of them. Once Dark Matter forms a gravity trough, baryonic gas accumulates like water in a bathtub. At the bottom of such troughs, galaxies of baryonic matter form. When enough baryonic matter has accumulated, stars finally form. Space between galaxies and stars is not empty, but full of Dark Matter, which is essential for the development of a galactic environment where conditions become suitable for the existence of life.

Stars convert the original hydrogen to helium, carbon, and oxygen. Massive stars burn it up further to iron. Stellar wind blows out the ashes into interstellar space. There, the heavy elements mix with original matter and form the next generation of stars. After some three or more generations, there are enough of the heavy elements to form Earth-like planets with solid surfaces, cores of iron, and atmospheres containing oxygen.

It is not a coincidence that we live long after the Big Bang. More than ten billion years are needed for a star like the sun to form and life to develop on Earth. For billions of years cosmic space has expanded, and today its observable size is 45.7 billion light-years in radius. The expansion and current size constitute cosmic space and are necessary conditions for the present state of the universe.

Was the purpose of the universe to create humans? Does space need us humans to have a meaning? I cannot think of cosmic space as the stage for humans like the paradigm of the Middle Ages. Being such a small part of it, this seems to be an outrageously anthropocentric view. Yet inversely, mankind could not have evolved in a smaller universe. The unimaginable size of the universe is necessary for our existence. Thus it is not without relation to us. It is not meaningless, since we have to accept that we absolutely depend on it.

Modern Concepts of Physical Space

It is conceivable that the region of origin of the background radiation in the early universe, which is at a distance of 45.7 billion light-years today, is not the end of the world. We cannot see behind it, but still may ask the question: How much may space continue beyond the horizon limited by the opacity of the early universe? There is a hint on this question coming from the theory of inflation, developed in the early 1980s. A phase of rapid expansion of the universe was proposed in 1981 by Alan Guth[3] to explain several cosmic enigmas. Most surprisingly, the universe appears similar in opposite directions. How can that be, as these regions cannot be causally connected? The inflation theory proposes that the universe had a phase transition shortly after the Big Bang releasing a great amount of energy. Space expanded exponentially and doubled its size every 10^{-34} seconds. The inflation speed of space reached a velocity larger than the speed of light. As space does not refer to particles and no information was transferred with a velocity exceeding the speed of light, the superluminous expansion does not contradict the laws of relativity. When the energy supply was exhausted, the universe was a factor of 10^{40} to 10^{50} times larger than before. Parts of the universe that are now on opposite sides of our horizon were in contact with each other and thus similar before inflation. Rapid expansion explains the uniformity of the universe in all directions. It also suggests that the universe is much larger than today's horizon at 45.7 billion light-years, defining the limit that we are able to observe. How much bigger is uncertain; the models disagree by more than 50 orders of magnitude and may be wrong altogether.

Expanding space is a new concept and deserves a closer look at the definition of space in physics. Isaac Newton (1643–1727) defined space as absolute, changeless, and unaffected by the physical processes taking place in it. He also assumed that space is Euclidian (i.e., not curved), three dimensional, and infinite. Ernst Mach (1838–1916) questioned these assumptions and postulated that the basic parameters of physics, including space, must not be defined by philosophical concepts, but by a method to measure them. This provided the basis for Albert Einstein's (1879–1955) theories of relativity. In practice, space is measured by the time it takes a particle or wave with known speed to go from point A to point B. Usually, light is used. If space is curved by the gravitation of matter, light travels along curved geodetic lines. In these new concepts of space, the galaxies are not expanding into a preexisting space, but space expands in three dimensions like the two-dimensional surface of a balloon that is blown up.

The definition of space by specifying a measurement method requires that at least in principle such a measurement is possible, or an equivalent observation can be made. Mach even stipulated the hypothesis that space is produced by matter and cannot exist without matter. No observation of a propagating wave before the Big Bang is known or even conceivable. Thus space and time before the Big Bang are not defined in physics.

The discovery of Albert Michelson (1881) and Edward Morley (1887) that the speed of light is independent of the motion of the observer required a modification of Newton's definition of space. Einstein coupled space and time in his Theory of Special Relativity to a four-dimensional entity. Space loses its absoluteness, but is relative to the reference system. The best-known consequence is the Lorentz contraction, according to which observers moving relative to each other measure different lengths for a given object. Thus we can choose the length of an object by moving it at a certain speed. Space becomes dependent on the observer. Moreover, in the Theory of General Relativity, space becomes dependent on the mass it includes. Its gravitation curves space. Thus the geometry of space is far from absolute, but determined by matter and the motion of the observer.

Modern string theories aim at unifying gravitation with the more basic physics of quantum mechanics. They add additional dimensions to space-time. These dimensions are not extended as the well-known three-dimensional space and time. The additional space dimensions are curved at the subatomic range, for instance into a closed loop. These theories are not confirmed and remain speculations until tested on experiments. Their final goal is to postulate space not as a given entity but as the result of basic forces and particles.

In conclusion, we should not see space as a preexisting frame of coordinates, into which the universe expands. Space is rather a part of the universe that originated, expands, and evolves with the cosmos.

Meaning of Cosmic Space

In 1977 Steven Weinberg shocked the world with his remark: "The more the universe seems comprehensible, the more it seems pointless."[4] Why did he say this? Pointless in this context means that the universe has no aim. It just happens to be here. The word pointless was translated in the German edition with *sinnlos* (meaningless), causing even more of an uproar. A pointless universe has no goal or plan, but finds its way by trial and error, chance and necessity. However, even if we consider the development of the universe to be in a haphazard manner, it still may possess meaning.

Meaning is a notion that I encounter most pressing in the connotation of the "meaning of life." The question has become a besieging one in recent times, in particular for modern individuals in Western society. In ancient times and in many societies still today, the meaning of one's life is to be part of a community, a nation, a tribe, or a clan. The fulfillment of life is met with fulfilling one's duty as a part of a larger unit. In today's Western culture the point of life is often found in personal relations or in love. These could be one's relations with a spouse, family, friends, partners, or God. Such relations yield a role and orientation in life.

What could signify the meaning of the universe or the meaning of space? Obviously the answer cannot come from physics. I cannot imagine an equation where *M*, standing for meaning of space, is the result of some parameters and constants. And the result of the equation then would be a number, such as 42.[5] As in the example of the meaning of life, the meaning of space must be seen in a relation to a superior entity or to the whole of the universe. An example of such a relation is the medieval interpretation of the universe as a theater stage for human life. Here the relation is given by stage and actors. They fit to each other according to the play.

Obviously, the medieval cosmos as a stage is a metaphoric interpretation. The important part was to establish a relation between mankind and cosmos. As in the biblical view, humanity's duty was to praise God. The universe participated in this function and thus had a meaning. It was a necessary agent allowing mankind to achieve its goal. Today humanity has lost its undisputed importance and central location in the cosmos. The current conceptions suggest the opposite: cosmic evolution is the grand play and may be the ultimate goal, with mankind taking part in it. What then could be the meaning of the universe and of space? Can galaxies, stars, and planets take the role of humans and praise God? That is what some biblical texts do indeed suggest, most clearly, in the Psalms.[6] What do they want to say? Do these images still appeal and can they interpret the findings of astrophysics?

Here we encounter interpretations different from scientific explanations. To discuss metaphoric descriptions of meaning and in particular the meaning of space, we have to turn now to methods and types of interpretation, and thus to the different epistemologies of science and religion.

Explanation and Interpretation

Scientific theories explain the relation of cause and effect or, alternatively, they give the probability that a certain event will occur. The ideal form of such an explanation is the mathematical equation or a probability. Mathematical symbols are not to be confused with metaphors. Take as an example the most famous equation of physics: $E = m c^2$. If the mass m is doubled, so is the energy E. If c were zero, so would be the energy. Other forms of energy may be added to the right-hand side of the equation. Einstein's formula can be integrated over a distribution of mass density. The equation can be mathematically manipulated in many ways and still make a valid statement about reality.

Metaphors cannot be added or integrated. A metaphoric interpretation of the kind "the universe is the stage of human life" cannot be expressed mathematically. If the stage is enlarged a billion times, human size is not. However, in such a metaphoric scenario, stage and humans keep the relation to each other even if the stage is enlarged. As long as humans live in this universe and have some relation to it, the imagery conserves the relation in space and time. Thus metaphoric interpretations cannot be disproved. A metaphor is not an equation and *vice versa*.

Metaphoric interpretations do not further scientific explanations. The Newtonian explanation of planetary motions did not become more precise nor was changed at all by interpreting the solar system as a stage for humans. No scientific gap of the time, such as the formation of planets, was filled. The stage interpretation was still used after Newton, as it gave orientation to the details of scientific explanations. Metaphoric interpretation looks at scientific results from some distance and gives them a perspective.

Yet there are other images that offer different interpretations. The seventeenth-century Deists interpreted the universe as a clock. The clock metaphor originated under the impression of Newtonian mechanics of the solar system in the age of the rising mechanical industry. Planets circle the Sun in orbits that can be predicted years ahead with stupendous precision from the equations of motion including inertia and gravity. The original, the mechanical clockwork, operates automatically according to its design. It is completely determined by its cogwheels. If all of them are known, the clockwork is fully explained and understood. In a clockwork each cogwheel is necessary and thus has a meaning. The whole clock has a meaning as the clockmaker must have had a purpose to create it. If the clockwork is the basic pattern that is discovered similarly in other entities; it becomes an image. The mechanical clockwork interprets metaphorically through its characteristic function, behaving deterministically. The metaphor then suggests that the universe is predictable and rational, and has no room for randomness.

If a metaphor becomes a part of the worldview and is applied universally, it becomes a paradigm. In the eighteenth century the clockwork metaphor of the cosmos reached paradigm status, suggesting that even humans were machines.[7] The metaphor of the clockwork survives until modern times in the form of the interpretation of the universe as an enormous computer.[8] This image, however, suffers some mismatch in view of modern astrophysics and biology, describing the universe as a chaotic development where chance plays an important role. The postulation of an intelligent design also faces the same problem.

To believe that the universe is a divine creation is another interpretation. All things, including space and time, then have their existence from a common principle. In Judeo-Christian tradition the creative power is ultimately transcendent and not part of this universe, neither in space nor in time. From the biblical creation stories in Genesis 1 and 2, to the Psalms, from wisdom literature to the Prologue of John's Gospel, the creative force is described as caring and graceful. In accordance with cosmologies of their time and general assumption, the Old Testament creation stories put creation at the beginning of the universe.

Modern astrophysics and biology indicate that nothing in the present universe originated in the Big Bang. Everything, including galaxies, stars, and planets formed at a later stage. Nucleons, the building blocks of matter, came into existence after the beginning of time. Even space and time grow continuously. If the universe is interpreted as a creation, it must be a continuous process. In theology such an interpretation is known as *creatio continua*. Continuous creation of space is the interpretation I want to expand on in the following.

Meaningful Space

Does cosmic space have a purpose or even a meaning? This question cannot be answered by science, but depends on our interpretation of the universe and, even before that, of our own personal existence. Creation is one of the possible interpretations. From a rational point of view, meaning follows immediately assuming that the existence and evolution of the universe are God's will and for his glory. The absolute aim then defines a relation of all creatures to the absolute. We note however that this construction cannot be proven and appears arbitrary to many contemporaries.

Religious interpretations are empty constructs if not based on experience of some kind. Four centuries of modern science have shown that the empirical basis of religion cannot be gaps in scientific understanding or a design claimed to be apparent in scientific results. Gaps are continuously closed by new results, while new ones open. A religious interpretation of the universe can only be grounded on religious perceptions, such as visions, mystic experiences, or biographic incidents. Religious perceptions occur in translucent moments when the world becomes transparent for a different reality.[9] Such transcendent experiences may happen under a starry night sky, but they are always related to a personal experience of a human being.

Humans experience creation not primarily in the confrontation with scientific results, but with our own existence. Where did we come from, and where are we going? At the age of some ten years we realize that we have a personality and can say: "I am me." The time, we can say, is finite. It is the rest of our lifetime. This is the time given to us; we have no right to demand more, but it is our most precious gift. It is the way I read the story of Adam. Being made of *adamah* (dirt from the field) and given his biological body, God endowed him time to live. He also gave him space to live, namely an environment capable of life, and a partner. This is the myth of human adolescence for both men and women: finding identity, falling in love, becoming independent, making mistakes, et cetera. Everything starts with the initial gift of a lifetime and a space to live, the primary creation.

The gift of existence in time and space is a basic human experience. It is evident that some spatial conditions for life, such as fertile land, water, fruits, and livestock, must also be met, as told in the story of Adam. These conditions are also experienced as gifts and thus are part of the creation story. Probably none of the more than 2,500 planets currently known is fit for life. A planet very much like Earth is necessary, having a dense atmosphere containing oxygen, oceans of liquid water, fertile land, a magnetic shield against cosmic rays, and much more. Earth-like planets consist of heavy elements, ashes produced in previous generations of stars. Stars form from molecular gas contained in galaxies. Galaxies form in gravitational troughs produced by Dark Matter that has five times more total mass than baryonic matter and thus controls gravity on large scales. These large troughs form only in a universe expanding from a hot and dense state after the Big Bang. The expansion continues and even accelerates due to Dark Energy. It has bloated space to at least 45.7 billion light-years, the size of today's observable universe. Humanity could not have evolved in a much smaller universe. Cosmic dimensions of space are needed for life.

Being the hatchery of humanity cannot anymore be the meaning of the universe. The size of the scientifically explored universe has grown dramatically since the Middle Ages. The human species is a minuscule part of the evolving universe at a very particular place and time. If space has a meaning, then it is the same as for the universe including humanity.

If human existence is interpreted as a gift, endowed by a Creator, the universe may also logically be interpreted as a gift like Adam's garden, as it is necessary for our being. Space then loses its frightening largeness, but becomes part of everybody's personal gift. A gift is in relation to the giver and the presentee. In this relation, space obtains a meaning.

The notion of creation implies a different quality when used for interpretation instead of explanation. According to a text by Hans Weder,[10] we perceive as creation something pivotal that we recognize, astonished, as a gift. This may serve as a definition for a modern understanding of creation. Creation is not directly noticeable in scientific results, but is recognized in a particular instant of personal amazement.

Can we perceive space as a creation? We may be amazed at how big it is, yet being related to us. We have ideas about the limits of space and understand its nature enshrouded in space-time, but we totally depend on it, and our becoming would not be conceivable without it. Amazed, we may notice that space grows out of nothing and the world could not exist otherwise. Considering that this growth is beyond our control, and at present even beyond our understanding, I can perceive it as given. Thus cosmic space may be interpreted as creation according to the above definition. This is an interpretation that is not coercive; a person may choose not to be amazed and ignore the character of the gift. Yet I am overwhelmed when realizing that my little person is part of this awesome universe and would not be possible without expanding space.[11]

Conclusions

Today's astrophysics describes the universe as a widely coupled system in evolution. Space, expanding since the Big Bang and as large as observed today, is a necessary ingredient for this cosmic unfolding. In this new view of the universe, Blaise Pascal should not be afraid anymore of the vastness of space. Cosmic space may be interpreted as part of the gift of life, necessary for a universe habitable for humans. The gift interpretation is the essence of a modern concept of creation, exemplified already in the creation story of Adam. Metaphoric interpretations need to be tested on their value as paradigms of a worldview. In my opinion, continuous divine creation of space is not only a possible interpretation in views of scientific results, but is consistent and appropriate. It fits the currently known astronomical universe at least as well as others such as the clockwork, computer, or intelligent design.

The notion of creation yields meaning through the relation between Creator and creatures. It must be stressed again that in this definition creation cannot be just a dogma added to scientific results but must be related to existential experiences. Creation is based on the perception of life and space as gifts. Belief in creation is not endangered by scientific resultsbut by neglecting and forgetting the basis of this experience. Personal perceptions, such as amazement, are nexuses between space in science and space in religion.

Notes

1. Michael Hoskin, *The Cambridge Concise History of Astronomy* (Cambridge: Cambridge University Press, 1999), 94.
2. "Le silence éternel de ces espaces infinis m'effraie." Blaise Pascal, *Pensées.* English translation by W. F. Trotter, http://oregonstate.edu/instruct/phl302/texts/pascal/pensees-contents.html.
3. Alan H. Guth presented his ideas for the general public in *The Inflationary Universe: The Quest for a New Theory of Cosmic Origins* (Reading, MA: Addison-Wesley, 1997).
4. Steven Weinberg, *The First Three Minutes* (New York: Basic Books, 1977), 154.
5. In Douglas Adam's 1979 science-fiction novel *A Hitch Hiker's Guide to the Galaxy* (London: Pan Books), a computer was set up to calculate the "answer to the ultimate question of life, the universe, and everything." The result was 42 to the base 13, but the question was forgotten.
6. Psalm 19:1: "The heavens declare the glory of God, and the sky above proclaims his handiwork." ESV Bible online.
7. The clockwork interpretation of the universe was applied to humans by Julien Offray de *La Mettrie* (1709–1751) in his book *L'homme machine* (anonymous 1748).
8. Konrad Zuse, the early pioneer of computers, proposed in the 1940s that the cosmos was a huge calculator (Konrad Zuse, *Rechnender Raum,* [Braunschweig: Vieweg, 1969]). More recently Seth Lloyd developed the hypothesis that the universe was basically working like a quantum computer. Its program consists in generating the whole universe including mankind. With the complete knowledge of all physical laws it were therefore possible to develop a quantum computer modeling exhaustively the universe. We would then be able to explain and completely understand the universe. Lloyd's ideas appeared in *Programming the Universe: A Quantum Computer Scientist Takes on the Cosmos* (New York: Alfred A. Knopf, 2006).
9. John Gatta describes nature becoming icon-like in a moment of devotion, in *Making Nature Sacred* (Oxford: Oxford University Press, 2004), 242.
10. Based on a text by Hans Weder, in *Kosmologie und Kreativität* (S. 68; Leipzig: Evangelische Verlagsanstalt, 1999).
11. A comprehensive exposition of the gift metaphor of creation was published by the author in *Das geschenkte Universum—Astrophysik und Schöpfung*, 2nd ed. (Düsseldorf: Patmos, 2010); English translation, *The Given Universe*, forthcoming in 2015.

Chapter 3

Faith and the Map: On the Metaphysical Nature of Visual Spatial Representation

Franco Farinelli

Why do we blindly believe in maps? It is generally thought that this is such a simple question that nobody has seriously considered asking it. However, this is also a question that touches on the most hidden recesses of Western culture, and it is a very long history, one that begins with Plato.

Let us take that complex passage wherein Plato explains how God shaped the soul by mixing up three different kinds of realities (Being, Sameness, and Difference), in order to produce one single Form.[1] It is worth highlighting that, according to Plato, the process takes place in such a way that a synthetical and mediating reality *par excellence* is shaped: the most conspicuous one, the one that makes up the intermediate sphere between the sensible and the intelligible. It is this passage that resounds in Serres's pages where the latter opposes the automorphism of the baker, who starts with a square dough and always comes back, with his manipulation, to the original square form, to the ability of the woman baker who, also by constantly cutting, imparts a Brownian movement to all points that compose the dough, thus including in it its time, circumstances, and fluctuations.[2] The action of the Demiurge is closer to the woman baker's than to the baker's, and the final result, the "world soul" that hierarchically preexists the world and therefore governs it, is a whole made up of many factors that are all intermediate beings between the ideal and the phenomenal.[3] Moreover, divine manipulation imposes on the world soul a capacity to move in a harmonic way, which is governed by numbers, for it takes away some parts of the initial mixture in accordance with proportional intervals and it fills them up with bonds of the same nature, so that it reconfigures the whole original mixture. But the decisive operation is yet to be accomplished. Having done all this, the Demiurge divides into two parts all the mixture along its length, then he juxtaposes them so as to form an X, unites each bar with itself and subsequently the two bars with each other in the point opposite the first intersection; finally, he wraps them round so as to form two circles, an internal and an external one.

Plato's description goes on, because the process is not accomplished yet, but we need not go on following him on this. For us what is most important, instead, is the fact of recognizing in the Demiurge's work the model of the squaring of a sheet, a practice without which no geometrical sign could exist, as it is such squaring that allows the tabula rasa to receive it. The tablet, the map, could hardly contain the uncertain traces of our free hand, its trembling expression, let alone the more imperious sign imposed by a set square or by compasses, without the intervention of such a ritual. Where does the *X* come from, the crucial chiasmus that God entrusts with the decisive movement for the construction of the world soul, if not from the two diagonals that we first have to trace on the paper in order to move from ornate drawing to geometrical drawing, thus founding and establishing the geographical image of the world? In other words, the map (the *copula mundi* of the Renaissance Neoplatonists)[4] is from the beginning the bridge between the metaphysical and the physical world. And in medieval times the map acts as the bridge between the world of Divinity and the world of Mankind, which is the reason why we still blindly believe in them.

Bodies

In his analysis of the "double body" of the king—which refers to the medieval idea that the king has an invisible and immortal body along with a mortal one—Ernst Kantorowicz refers to the corporative doctrine of the Catholic Church, which became dogma with Boniface VIII in 1302. As can be gathered in the bull *Unam Sanctam*, the church "represents a mystical body, with Christ at its head, and with God at the head of Christ." This idea contradicts what, up to the middle of the twelfth century, had been professed by the Roman Church, namely that the mystical body of Christ was the Eucharist.

In 1215, because of the doctrines of Berengario de Tours and the heretical sects, which tended to spiritualize excessively the sacrament of the altar, the pope was forced to strongly reaffirm the real presence of both the human and divine natures of Christ in the Eucharist.

Through the dogma of transubstantiation the consecrated bread is therefore defined *corpus verum*, or *corpus naturale*, or simply *corpus Christi*. As a consequence, this expression, which was used by Paul to designate the Christian church, came to designate the Eucharist. Conversely, and at the same time, the expression "mystical body," whose origin was liturgical and sacramental, started to designate the body of Christian society united in the sacrament of the altar and acquired a "sociological" meaning. As Kantorowicz concludes, the new theorization of the "two bodies of the Lord" (the individual *corpus verum* of the altar and the collective *corpus mysticum* of the believers' community) is the most exact precursor of the theory of the "two bodies of the King."[5]

In Henri de Lubac's study on the nature of the mystical body that Kantorowicz himself refers to, things appear to be more complicated, because this expression involves, more or less directly, not two but three bodies: the body of Christ consecrated by the priest on the altar, that is to say the Eucharistic body; the blood and flesh of the passion, that is to say Christ on the cross; and, starting from the thirteenth century, the ecclesiastical body of the believers' community and the church. It is impossible here to sum up the very complex relations among these, the meaning they acquired, and the way they unfolded throughout the medieval period. However, two things must be remarked if we want to understand modernity. The first concerns the three different functions that these three meanings have, although such differences never impinge on their deeply rooted identity. These are: presence, memory of the sacrifice of which the Eucharist is a method of commemoration, and anticipation—the mystical body, the body of Christ received in communion stands for the announcement of his death. The second point is that, from the early years of Christianity, the idea of the church coincided with that of *ecumene*, that is, the body of Christianity as the body of the whole universe, or better, it coincided with the whole known and inhabited lands.[6] Altogether these two facts mean that what Kantorowicz saw as sociological meaning is in truth a geographical meaning and also point to the mystery surrounding our absolute faith on cartographic images. Or to put it better: it is the very reality, including its sociological form, which reveals its dependency on geographical representations, as will be illustrated here.

Had it not been for Augustine perhaps, we would not be here to talk about Paradise. At the beginning, in the Bible, there was an ambiguous Hebrew term, *miqedem*. In his *Vulgata*, the authoritative canonical Latin translation adopted by Christianity in the West throughout the medieval period, Jerome interprets this word in relation to its temporal meaning as "at the beginning." This reading contrasts with those of other versions of the Bible—for example the Jerusalem Bible—in which this term has a spatial meaning; it means "to the East." In any case in Genesis, the term in question refers to Paradise, the watered garden in bloom, full of trees and flowers surrounded by walls and populated by animals in which the Lord put man so that he could keep and cultivate it.

The Body of Christ

It was Augustine who insisted on a precisely literal rather than allegorical interpretation of the story of Paradise. He wanted Adam to be a creature in flesh and bone living in an equally real garden, and like all real things, was in turn part of the physical world.[7] In short, what Augustine referred to was something that could have a place on a map. By map, we here mean a representation without which the very idea of the world as something unique, coherent, and homogeneous could hardly exist, as the cartographic representation of Paradise also shows. It is precisely the paradoxical task of tracing on world maps a place that was not of this world that justified the reduction of the whole earth to a map, as if in Paradise itself—understood as a sort of vehicle and agent of this reduction—was hidden the mystery of mysteries, that is to say the mystery of the equivalence between the world and its cartographic image.

As Hugh of Saint-Victor wrote in the first half of the twelfth century, "all sensible world is a sort of book written by God's finger."[8] And for Ugo, author of a famous *Descriptio mappae mundi*, the book par excellence was the map.[9] Years ago, Hans Blumenberg drew attention to this overlooked metaphor, noting that the foundation of great modern narrative on the nature of evolution was a model based on the act of unrolling (*evolutio*) the scroll.[10] This has been so up until the first century CE, when the scroll was substituted by the codex, which consisted of sheets of paper placed one upon the other and not, as it was before, one after the other. Blumenberg's conclusion highlighted that what appeared to be close together in linguistics was in fact comparatively distant when we think of the time that separates Darwin from the beginning of our era. This comparative reflection is valid for the influence of maps and their forms on the contemporary world image.

Differently from what happens with language, on a map no term can be ambiguous: the beginning and the Orient are the same thing.

As a matter of fact, almost up until the thirteenth century, Christian world maps—from the small ones that illustrate commentaries to classic texts to the huge ones that were hung on walls of churches—show the East high up and the garden of Eden at its top (or almost), with few exceptions. This cartographic temporal and spatial coincidence and the presence of Paradise show that throughout the Middle Ages all maps were a system that could represent not just space and time but also past, present, and future. The map, in fact, did not just show an inventory of what existed, but also represented the nature of the real and, at the same time, provided a prognosis that consisted of the universal message of Christianity and the divine plan for the salvation of humanity. Whether we are believers or not, all this still influences our relationship with the world and determines its mechanisms in a way that is powerful and of which we are barely aware. The dynamic of this relation—one that is cartographic—is cogent because it mirrors the threefold coexistence of memory, presence, and anticipation that characterizes the mystery of the Eucharist.

Let us consider for example what was once the gigantic map by Erbstorf as we can see it today in its facsimile made after it was destroyed during the Allies' Hannover bombing.[11] It is a huge Christ, whose head is placed right next to an illustration of Paradise that shows the scene of diabolical temptation. The body of Christ mingles with the Earth, which he embraces so that the image of the latter becomes a gigantic Host. This globe was made between 1235 and 1240, after the fierce controversy between Berengar of Tours and the Roman Catholic Church concerning the real presence, during mass and after the consecration, of Christ's true body and blood, a presence that, according to Berengario, was not real but merely symbolic. As a consequence, after the Eucharist dispute the diffusion of maps similar to that of Erbstorf implied an evident transfer: in the same way in which the host became the true body of Christ through the Eucharist, so the map, the host of the Earth, became its true body.

How would it be possible otherwise to explain the blind faith we have in maps, a faith that is widespread though it has remained up to now unconscious and represents one of the distinctive traits of modernity? Precisely because Columbus was the first to trust blindly in maps, he is also the first modern traveler.[12] For Columbus, however, maps remained a prophecy, as they were throughout the medieval period. It is for this reason that when Columbus saw the mouth of the Orinoco he was led into thinking that he had arrived near the four great rivers of Eden.[13]

The Body of Europe

It is precisely the disappearance of Paradise from maps during the modern period that shows the end of the awareness concerning the prophetic, or at least prognostic, nature of geographic images. For us, as it was for the less disingenuous medieval men and women, these images continue to prefigure what is going to happen, even though we know well that they merely register what is there. The decisive move took place during the middle of the fifteenth century when Fra Mauro, the monk who worked in the silence of the monastery of San Michele in Murano, completed the colored circular map that vividly marks the passage from the medieval to the modern image of the world.[14] This transition is constituted by the following: for the first time Paradise is placed in the corner. While it is still included on the map, it falls outside the circle that marks the boundaries of the Earth. In this way, it suffers the same punishment that God inflicted on Adam and Eve when he banished them from Eden. With the banishment of the original event of humanity from the representation of the Earth, this map is no more able to host, as previously, places that are at the same time events. The temporal dimension (*saeculum*) of reality and the spatial one (*mundus*) for the first time are systematically distinguished from each other.[15] So time on the map disappears and only space remains and becomes the sole authority. Space appears in its Ptolemaic form, that is to say geometrical and functional to the construction of the new world, the world we live in.

If the banishment of the first couple from Eden marked the beginning of history, the banishment of Eden from maps marks the beginning of what in the Western world is called geography. In geography, as well as in the needs and aspirations of all descendants of Adam and Eve, Paradise still exists in the form of archetype and prototype. Nowadays Paradise does not correspond to a material sign, but to an immaterial model that is in everybody's imagination: the idea of place, of that ambit that, as Doreen Massey explained, is not known unless you inhabit it, or you are inside it.[16] Or, paraphrasing what Proust wrote about Paradise, that place that is true and unique because we have lost it and since then we continue to look for it.

We can now understand why Novalis in 1799 defined Christian Europe as a "large spiritual Kingdom" similar to a "continent with a human form"; he had a visual or mental image of a medieval *mappa mundi*, one of those which used to adorn cathedral walls or prayer books and in which Christ and the Earth were one body.[17] We can also guess the reason why during the Medieval period the church was the most advanced institution for territorial organization,[18] so much so that in the modern period its code was to be used as the model in controversies concerning boundaries.[19]

Novalis's Europe was still that Christendom, believed to be able to stop the advent of the Antichrist, that is to say the end of the world. Outside Europe, pagans' territories were "mission lands" that were allocated by the pope to some Christian Prince or the other. Islamic territories were instead the land of enemies, subjected to those expeditions of conquest known as the Crusades or the Holy Wars. The European territory—which was apparently in the process of expanding but also, and at the same time, forced to contract—was divided according to the law of that period into several institutional subjects: principalities, crowns, churches, cloisters, monasteries, dominions, castles, neighborhoods and communities, and *universitates* of all kinds.[20] The European territory was like the mystical body on the altar; this was not just a body born from the Virgin but also the true body of Christ, the body in flesh and blood sacrificed for redemption. According to the well-known assimilation between the holy bread and the bread of life, breaking the holy bread (that is to say the terrestrial body) coincided with the revelation of the divine word. By breaking the holy bread, the revelation of mystery revealed itself. Hence, writing can be considered as the bread; like the bread it has a crust and a soft part; the crust is literal meaning, the soft part is intellect (*intellectus*). Intellect is hidden by literal meaning and hence can be revealed and thoroughly understood only by breaking and removing the crust.[21]

The Body of Utopia

The close analogy between the act of breaking bread and what Peirce wrote on the nature of the sign is clear. The table of the Eucharist, in the case of the Earth, the material foundation of the map, can be explicitly and exactly seen as the "quasi-interpreter," that is to say as the "quasi-mind."[22] However the awareness of this analogy does not hold when faced with the elimination of Paradise from the terrestrial space, intended here as the world that can be inhabited and known. In fact, such an elimination brings forth the elimination from geographical representations of all references to past and future, to memory and prophecy; it is an elimination that transforms these representations in pure and mere presences.[23]

From an ontological point of view, these presences are the first products of a reductive cartographic representation of reality. All these produce a cognitive disconnection between the two "quasi-minds" Peirce talked about; the one that functions as "quasi-interpreter" does not recognize the other one—the material one—as co-constitutive of the sign, and hence ends up ignoring it completely. From this perspective, the idea of Utopia is one of the last manifestations of the awareness of such a collaboration, the removal of which, however, does not bring forth the end of the functioning of the Table. The name Thomas More gave to Utopia was Nusquama.[24] Utopia, or the country that according his own name does not exist, provides a way to understand how at the beginning of modernity there was still an awareness of the map as mind. This is the secret behind the foundation of the ambiguous country of Utopia and provides an explanation about how this country can exist and not exist at the same time, how it can be simultaneously true and untrue, as mentioned in the brief poem in Latin that appears in the verso of the second page of the manuscript.[25] Here in four verses it says that Utopia in the beginning was not an island and it was Utopus, its Master, that made it so; it also says that Utopia is the only land that represents for men the philosophical city, though it has nothing philosophical in it, and that Utopia gladly gives away all that it possesses and accepts the best of things from others.

As Raphael Hythloday, the Portuguese sailor and narrator in Utopia, says, in the beginning this island was a peninsula, that is to say a quasi-island. In other words, there was something that was very concrete and not at all idea. In order to become what it is, Utopia needed just a small change. Through this change, what is inside meets with what is outside; this refers to an encounter between the two "quasi-minds" Peirce talks about. The activity of one consists in the concession—that is to say the emission—of the internal datum, the activity of the other consists in the reception—that is to say in the interpretation of the external datum. The riddle consists in trying to find out the vehicle, or instrument, that allows the two to be welded together, as Peirce has it. In this respect Raphael's tales, and in particular the language that makes up the history and geography of the island, are revealing. The first name of Utopia was Abraxa, which means "on which it does not rain"; its Prince is called Ademus that is to say "without people"; the capital is Amaurote that means "barely visible" but could also be interpreted as "something that can be erased"; southeast of Utopia is the country of the Achoriens, a name that translates as "without region"; and finally—though the list could continue—the main river that crosses the island is called Anyder, which literally means "without water."

The name of the river is significant because it sheds light on all the other names and shows the affinities these share. Suffice here to recall *The Republic*—one of the few (if not the only) Platonic dialogues known during the medieval period—and in particular its grandiose and extraordinary final part.[26] Here Er, another pilgrim who travels beyond the boundaries of reality, recounts his journey to the scorching plain of Forgetfulness along the subterranean path that the dead have to go through if they want to drink the water from the river of forgetfulness and reincarnate.[27] In these plains are rivers but no vessel can hold their water, there are trees but none of them provides shade. Up until some years back, young children born in what was once the Magna Graecia knew the answer to this riddle—the same one that concerns Utopia—from a very young age: the country where there are rivers whose water no vessel can hold and where the trees do not shade is a country that is drawn on paper and for this reason, like Utopia, is everywhere and nowhere.[28]

When Pierre Gilles says to his friend Busleyden that More's book provides its readers with an image of the island as if this was drawn or painted, the word "painted" should be understood literally.[29] In one of his letters, More explains to Gilles that names of places affirm their historical reality precisely because their meaning corresponds to nothing, *nihil significantia*. Louis Marin translates this as "they made no sense at all" and misinterprets More.[30] According to More, in fact, these names mean nothingness, that is to say they do not refer to *realia*, but they are signs. It is precisely as signs that names are endowed with historical determinacy and a modern nature. To use Heidegger's words, names provide proof of the fact that they belong to the "era of the image of world," that is to say the era of the reduction of the world to a map.[31] Such a reduction has been made possible by our absolute belief in maps. We blindly believe in maps because in each one of them is hidden and silently beats the heart of the "sacrament of the sacraments," the mystery of mysteries of the Western world. We believe in maps because each of them is the Eucharist of the body of the Earth, the altar of its sacrifice in front of which we continue to renew our unaware act of faith. It is not by chance that at the beginning of the nineteenth century, Carl Ritter denounced the "cartographic dictatorship" of geographical discourse only after he had reminded his readers that all scientists are bound to declare the nature of their subjective beliefs. In Ritter's case this meant a declaration of his Christian faith.[32] This is proof of the fact that at that time and in critical geographical discourse one can find a persisting awareness—albeit silent and implicit—that the objective and scientific image of the world had first and foremost a sacred character.

Notes

1. Plato, *Timaeus*, 34 b 10–36 c 6 (London-Leiden: Brill, 1975).
2. Michel Serres, *Les origines de la géometrie* (Paris: Flammarion, 1993), 88–91.
3. Giovanni Reale, *Per una nuova interpretazione di Platone. Rilettura della metafisica dei grandi dialoghi alla luce delle "Dottrine non scritte"* (Milano: Vita e Pensiero, 1997), 660–61; Harold Cherniss, *Aristotle's Criticism of Plato and the Academy*, vol. 1 (New York: Praeger, 1962), 409, n. 337.
4. Giovanni Reale, *Platone. Alla ricerca della sapienza segreta* (Milano: Rizzoli, 1998), 261.
5. Ernst Kantorowicz, *The King's Two Bodies: A Study in Medieval Political Theology* (Princeton, NJ: Princeton University Press, 1957), 167–71.
6. Henri de Lubac, *Corpus Mysticum. L'Eucharistie et l'Église au Moyen Age* (Paris: Aubier-Montaigne, 1949), 83–106.
7. Alessandro Scafi, *Mapping Paradise. A History of Heaven on Earth* (London: The British Library, 2006), 22–35, 57–58.
8. Hugh of Saint-Victor, *Eruditio Didascalica*, in *Opera Omnia*, II. (Paris: Migne, 1880), coll. 739–838.
9. Patrick Gautier Dalché, *La "Descriptio Mappae Mundi" de Hugues de Saint-Victor. Texte inédite avec introduction et commentaire* (Paris: Études Augustiniennes, 1988).
10. Hans Blumenberg, *Die Lesbarkeit der Welt* (Frankfurt a.M.: Suhrkamp, 1981), 13.
11. http://www.ebstorfer-weltkarte.de (December 23, 2013); Birgit Hahn-Woernle, *Die Erbstorfer Weltkarte* (Ebstorf: Kloster, 1987).
12. Franco Farinelli, *Geografia. Un'introduzione ai modelli del mondo* (Torino: Einaudi, 2003), 18–19.
13. Consuelo Varela, ed., *Cristóbal Colón. Textos y documentos completos* (Madrid: Alianza Editorial, 1982), 217–23.
14. Piero Falchetta, *Fra Mauro's World Map* (Turnhout: Brepols, 2006).
15. Scafi, *Mapping*, 101–4, 201–6.
16. Doreen Massey, Jess Massey, and Pat Jess, eds., *A Place in the World? Places, Cultures and Globalization* (Oxford: Oxford University Press, 1995), 49–50.
17. Novalis, *Die Christenheit oder Europa* (Stuttgart: Reclam, 1962), 35.
18. Robert Sack, *Human Territoriality: Its Theory and History* (Cambridge: Cambridge University Press, 1986), 92.
19. Paolo Marchetti, *De Iure Finium. Diritto e confini tra tardo medioevo ed età moderna* (Milano: Giuffré, 2001), 98–99.

20. Carl Schmitt, *Der Nomos der Erde in Völkerrecht des Jus Publicum Europaeum* (Berlin: Duncker u. Humblot, 1974), 42–44.
21. De Lubac, *Corpus Mysticum*, 99–101.
22. According to Peirce, "any set of Signs which are so connected that a complex of two of them can have one interpretant, must be Determinations of one Sign which is a Quasi-mind," which means that, admitting "that connected Signs must have a Quasi-mind, it may further be declared that there can be no isolated sign. Moreover, signs require at least two Quasi-minds; a Quasi-utterer and a Quasi-interpreter; and although these two are at one (i.e., are one mind) in the sign itself, they must be nevertheless be distinct. In the Sign the are, so to say, welded." Charles Sanders Peirce, *Collected Papers*, Charles Hartshorne and Paul Weiss, eds. (Cambridge, MA: The Belknap Press of Harvard University Press, 1965–66), 438–39.
23. Franco Farinelli *I segni del mondo. Immagine cartografica e discorso geografico in età moderna* (Firenze: La Nuova Italia, 1992), 11.
24. Germain Marc'hadour and Roland Galicois, eds., *Erasme de Rotterdam et Thomas More: Correspondance* (Sherbrooke: Université de Sherbrooke, 1985), 33.
25. Thomas More, *Complete Works*, IV (New-Haven, CT and London: Yale University Press, 1965), 251.
26. Plato, *The Republic* (Oxford: Oxford University Press, 1993), 619–21.
27. Jean-Pierre Vernant, *Mythe et pensée chez les Grecs. Études de psychologie historique* (Paris: Maspero, 1966), 125–43.
28. Franco Farinelli, *L'invenzione della Terra* (Palermo: Sellerio, 2007), 98–105.
29. More, *Works*, 21.
30. Louis Marin, *Utopiques: Jeux d'espaces* (Paris: Les Éditions de Minuit, 1973), 121.
31. Farinelli, *I segni del mondo*, 55–70.
32. Carl Ritter, *Einleitung zur allgemeinen vergleichenden Geographie, und Abhandlungen zur Begründung einer mehr wissenschaftlichen Behandlung der Erdkunde* (Berlin: Reimer, 1852), 26–27, 33–35.

Chapter 4

Human Dwelling: A Philosophical Question Concerning Place and Space*

Carla Danani

The lack of consideration given to spatiality in dealing with human reality is neither a contemporary failing nor the product of globalization or the spread of new forms of virtual reality. We often fail to acknowledge that human beings, and therefore their actions and thoughts, relate to each other and to their surrounding environments within the "medium" of space: this means that we are always localized.[1] We shall use the terms space and place interchangeably in this article, although we do not regard them as synonyms. A place is a marked space, a space that is qualitatively defined, that possesses an imprint, a character, a limit, and a history.

Phone calls, instant messaging, and video-conferencing seem to make the "here" extremely fluid and more indefinable than the "when." This is unsurprising, as nowadays when we make a phone call we are more likely to ask, "Where are you" rather than, "Who's calling?" because the caller's name usually appears on our contacts' list, whereas their location is unspecified (in the absence of a tracking device). However, the "where" is clearly not a matter of indifference to us, given the ever-growing demand for such devices.

Yet, on closer inspection human beings seem to be more familiar with space than with time. We often see people wearing a watch but we rarely see people with a compass. We always need to know the time but we never ask ourselves where we are because we think we know. The question "Where am I?" denotes a situation of discomfort: either a recovery from a situation in which someone is not in full possession of his/her faculties (due to physiological causes or external constraints) or an expression of disappointment at being somewhere one does not want to be. We use devices such as GPS only to locate the places and routes that are unfamiliar to us, while we make regular use of tools that indicate time on a daily and continuous basis.

Moreover, in order to explain who we are, today we often say where we come from, we tell people that we are from Europe, or from Italy, or from a little town over there: it depends on how far we are from that place and how far the interlocutor is perceived to be from us. The expression "I'm from...," unlike saying "I come from," suggests that one's origin is experienced as belonging. Even if nowadays, in one lifetime, there are several places where we may live, there is always a particular place with which there is a stronger bond. Today one's origin might not be revealed by religion or skin color, but rather in the accent characterized by a residual dialect inflection, by the imagination full of images of sea or land, of lions and giraffes or ibexes, or by the metaphors inhabited by different figures.

Even though individuals and entire peoples move from one place to another, so that our "roots" have been replaced by anchors, to use Bauman's words, and even though the media offers us a shared world to inhabit, we were born and we live somewhere. Whether this "somewhere" can be defined as a cosmos, a world, an earth, a place or so on, existence always takes place in a "where." For this reason it is important to take spatiality into consideration when it comes to the philosophy of experience, ethics, and politics. It is an essential issue for anthropology and, therefore, for any human discussion.

Spatiality's importance is signaled by a dialogical interrelationship between its components: each responding to the other. The relationship between ethics and the experience of spatiality initiates the dialogue, as one considers the importance of this dimension for human life, and one wonders how to treat space in an ethical way and to look for ethical implications in the field of architecture, urban design, and environmental management. But should we not also pose to ourselves the opposite question: If human life is always transcendentally localized, how might it inform the creation of ethics for human beings? There also seems to be a direct connection with politics. In this instance too there is a dialogical movement of thought. Spatiality has to be considered both as the relevant dimension for living together and also as the horizon against which political thinking is shaped.

Space and Corporeality

If human existence happens in and through place, it is because it is essentially a life in and through the body. A human being is an embodied consciousness. This means that corporeality and consciousness exist as distinct but inextricably interconnected realities. The objectification of physicality is, therefore, always crossed by the subjective intentionality, and this, in turn, can only exist when it is objectified.[2]

Therefore, space is both the medium in which the perception of objects can take place and also the medium in which one can move and act. And since we are able to move and act, we have to assume a certain grasp on space, thanks to which we can orientate and localize ourselves.

Being able to orientate and localize ourselves requires a certain knowledge of the context and in particular a knowledge of ourselves within such a particular context. While we orientate ourselves, the certainty of our position creates the dizzying possibility of being somewhere else: this causes an awareness that this place is only "one" place among many others. This place, no matter how much we bond to it, does not coincide with ourselves. It is not only the uneasiness we sometimes experience in a place that constantly reminds us that there is something outside us. Even when we feel comfortable in a place, we are guests of this place, we perceive it as something different from us, something from which we are separated, while we still struggle to enjoy it, to find protection and concentration in it, to become familiar with it.[3] We are "here" but we do not coincide with our "here." We can leave this place and move somewhere else, and our "here" can become the "here" of other people.

Talking about Space

Talking about space is a difficult task. First of all, it is something that we experience. We act within it in a pre-reflexive way, without asking ourselves what we think about it, without being aware of it, on the basis of a familiarity that is more ancient than thought.[4]

The structure of space is complex, it resists easy categorizations or characterizations and it includes subjective and objective elements. Cassirer's *An Essay on Man* is often quoted with respect to his distinction of fundamentally different types of spatial and temporal experiences that cannot be put at the same level.[5] The lower level is defined as organic space and time. If we think about ants or bees, or the ability of chicks to orientate themselves and peck grains on the ground, we realize that a representation and an idea of space are not necessary to act in a place, and as a matter of fact these animals do not have a sense of spatial relationships. Organic space is also defined as "action space," where actions are usually determined by practical needs and interests. On the contrary, "perceptive space" is not an immediate fact given by senses, but it includes several experiential aspects, such as optical, acoustic, tactile, and kinesthetic elements. Perceptive space is the field of investigation of those who are interested in understanding the origins of perceptions. However, as Cassirer points out, from the point of view of a general theory of knowledge and of philosophical anthropology, rather than investigating the origin and development of perceptive space, it is necessary to analyze "symbolic space," which will lead us towards the boundary between the human and animal world. Cassirer points out that having a representation of something is completely different from just dealing with it. Knowing how to use something for practical purposes is not enough in order to achieve its representation. We also need to have a general idea of it and to consider it from different points of view, in order to discover its relationships with other objects. It is also necessary to localize it and define its place within a general system. You can be familiar with some aspects of experience without having a representation of them: you can know an area very well and be able to move around very easily, without being able to trace a map of it. In the history of culture, for instance, we had to acquire a high degree of generalization to achieve the conception of an ordered outer space. Therefore, space for Cassirer has an organic dimension, a perceptive dimension, and a symbolic dimension. In each case, space is the medium where we establish relationships with the world and with the plurality that inhabits the world. "Our space," "our place" is always somewhere where we get in touch with what is other from ourselves.

We can also try to distinguish between subjective and objective space. Subjective space is the grasp on space that is inseparable from a creature's sensory and cognitive capacities and motor skills, from its body and environmental awareness and from its ability to interact with the environment itself. In other words, it is the perspective of an active involvement. It is an experiential space that is also an activity space, structured and focused on the corporeality of the subject and varying according to the subject's motor, sensory, and cognitive abilities. Subjective space is the space of behavior and does not depend on a theory of space.

In contrast to commonly received opinion, Heidegger argues that we cannot derive the concept of objective space simply from the idea of subjective space, nor from mere connections between subjective spaces. The idea of such connections implies the idea of objective space as the space where we can have this localization regardless of the particularity of experience. We cannot even speak about objective space by eliminating the perspective elements, since we can eliminate the perspective elements only if we already possess the idea of an objective space, independent of subjective experiences.

As a consequence, we cannot derive objective space from subjective space. In the same way, subjective space cannot be derived from objective space: the idea of an experiential point of view is what constitutes the subjective space and is exactly what is missing in the idea of objective space.

The fact that we cannot derive one space from the other does not mean that they are not interconnected. Even though they are conceptually different, we cannot understand one without having an idea of the other. They are correlative: thinking in a radical way about one type of space implies possessing the idea of the other. The idea of subjective space requires a perspective within which this space is apprehended. Similarly, the idea of being in a perspective implies a transcendence that is made possible by the awareness of a wider context in which our point of view exists. And yet, such awareness always occurs within a subjective grasp.

We experience reality as localization in a subjective way. However, the place we occupy does not become private because it does not coincide with ourselves, but it remains persistently common and it can be experienced from other perspectives in different ways.

Dwelling and Subverting

It has been said that human existence is the existence of an embodied consciousness that lives in a place. We also say that things have their own place, that they are somewhere. However, human beings and things live in places in two different ways. As a matter of fact, human beings always live among the things that inhabit their lives, even when they withdraw into their inner world.[6] Indeed, we say that human beings dwell, they relate intrinsically to the place where they live, while things just stay in it. People dwell in places through a process of familiarization (by hanging pictures on the walls, putting photos on their desks, planting trees in their garden or a hedge to establish a boundary, etc.) and orientation: in this way existence always possesses some coordinates and its logic is always a topological one.

This "being in" a place is being in a "here," within a structure that has its mode of being as a horizon. Every "here" is placed in a fundamental connection with every "there," be they understood as earth or world (leaving aside the debate about Heidegger's use of the two terms). We enter the world from "here" and the world defines "here" in relation to "there," preventing "here" from closing into its tamed finiteness,[7] into its fictive self-sufficiency.

The point of view of place that coincides with the practice of familiarization equates to a point of view on the world that subverts every possible appropriation: because the world itself represents the inappropriable. As a consequence, we can define dwelling—which is the mode of existence of human beings—as a practice of familiarization with what, from time to time, is our place, within a world in which we remain strangers, since, as we said, no "where" coincides with the self.

If we think about our home, we realize that it is characterized by a complex dynamic. Even if it represents the most intimate and cozy place for a human being, we cannot simply remove what cannot be assimilated. Or better, succeeding in such an aim would result in the transformation of the house into a prison.[8] On closer inspection, the dynamics of familiar and foreign, private and common seem to constitute the logic of the space itself.

It is the dynamic itself that enables the existence of a sacred space (the use of the term "sacred" is not intended to sacralize the space as if it were the bearer of a divine quality, but it is used to indicate what pertains to the religious experience), where the restless tension can open up another level of experience, inverting the spatial dynamic of a house.

In this context the tension turns out to be on another level: it is not the world of the inappropriable and the inextricable that can in some way become more and more familiar and knowable. Instead, it can be defined as "outside the world." Sacred space is something that we or our ancestors have built, but that we inhabit only occasionally: that is when we let our time be the *kairos* in which we face God rather than the *chronos* that measures our existence in the world.[9] A sacred space is true when it causes a thrill of anticipation to arise in the visitor, and she or he perceives an indefinite and powerful force that orders: "Be silent, because this place is full of overflowing presence." Even though at first sight it can be compared to any other place, it possesses many peculiarities. A sacred space shows its power by overlapping the connection between the world and the borderland. It is a peculiar experience: only the relationship with the Absolute enables the possibility of being "on the border."

A sacred space can be so named only when it is not experienced as a shelter to protect yourself, with the guarantee of the Absolute, from the concerns of a world that fails to become your homeland. A sacred space is not meant to offer a permanent residence and its threshold does not divide the troubles of history from the happiness of the *eschaton*. On the contrary, it is the place of a fragile anticipation.[10]

In this experience, the fact of being in that specific place becomes secondary. Some places, such as secluded and pilgrimage places, have always been an example of such experience, with different nuances. This does not mean that the location loses its meaning. The possibility of subverting the topological feature suggests a redefinition of place: it is not only considered as a place where you are "in" but also "through-towards," in a tension towards a fulfillment. The inherency within the place shows the traits of the dynamic relationship.

From this comes an interesting consideration: the subversion of place discloses the truth of the place. The subversive loyalty to our "here" tells us the right way of dwelling: first of all as a contrast with all the possible "theres," and secondly as the relationship with what the "here" cannot encompass.

The Symbolic Dimension

Sacred space seems to guard and reveal, suggesting the mystery of topology. As a limit-situation, it is a hermeneutical place that shows the symbolic nature of all places. Such a symbolic nature is the necessary condition for the "fidelity of subversion." Places are physical elements that have a metaphysical reference, the world's sense of being. They carry out the re-cognition of the cross-reference in terms of allusion and memory. The symbol creates an increase of reality through a displacement of meaning that can be defined as semantic impertinence.[11] In other words, the meaning of a sentence arises from the failure of the literal interpretation of the statement. The symbol is the constant interference, the bewilderment, the calculated error, the disorder of the ordinary world that produces a thrill towards new possibilities. It addresses the whole human being, not only his or her intelligence.[12] The symbols communicate their messages even if our rationality is not aware of them. The word symbol comes from the Greek *sym-bállein*, which means to put or throw together. The symbol is not a sign whose referent is fully knowable by the senses or conceivable by the mind, because for the mind the signifier and the signified are in the same level of reality, in the field of experience of phenomenal reality. The symbol has a multipurpose semantic ability; it has different levels of reference, that is, an essential meaning and some reflected meanings, which are in some way grasped through the senses. The crucial rule is the heterotopic break.[13] Take for example the advertisement of a famous Italian company: everything focuses around a piece of pasta, a fusillo, that does not only represent the pasta (first meaning), but also recalls the symbolic universe of family relationships (second meaning), when the piece of pasta is subjected to the heterotopic semantic impertinence and it is found not in a pot, but in the father's jacket who is abroad and far from home. Obviously, the transition from the first to the second meaning does not imply the destruction of the level of signs. It should be noted, in fact, that the transition is possible only if the sign, the piece of pasta in this case, remains the same. In the case of sacred space or symbolic religious space, its sacrality (second meaning) does not exist unless there is a secular, ordinary space (first meaning), typical of every culture and age. In turn, in the ordinary place, which is the objectification of a need, we may also find an orientation in the overall horizon of reflected meanings.

The sacred place clearly expresses and, at the same time, opens up an understanding of what is missing, as the condition of what is possible, and this is precisely the path of symbolic consciousness. Symbolic consciousness lies in the particular, but it also reads the universal aspect, holding together difference and identity. Steps, for example, act as the access point to a sacred place and, at the same time, effect a defamiliarization with the surrounding space. They seem to tell us "Take off your shoes, this is sacred ground."[14] The sacred place has a clear performative function: to free what has been forgotten, to awaken the concern that our homeland is elsewhere. The traditional architectural elements that inscribe this symbolic directrix are, among others, orientation, size, and the inscription of the square (world) in the circle of the whole structure or dome (God). Therefore, while establishing a relationship, it underlines the difference and the disproportion. From this perspective, the logic of waste, of monumentality, of the unnecessarily audacious structure of worship buildings acquire a meaning: they must express the impossible on the basis of functionality; they bring with them the mad task of seeing the invisible.

If sacred space is an explicit declaration of semantic impertinence, every place implies in some way a symbolic dimension, that is, it consists first of all in what it is and, secondly, in both the act of referring and in the set of references to which it refers. In this the utopian element is inscribed: a totalizing intention (not totalitarian), which lies in the ability to highlight what is missing, in an open tension never closed-in on itself. In this respect we can quote two interpretations that must be read with concern as the risks of what utopia might become. The first one is Carl Schmitt's reading, who claimed the nihilistic and deterritorialized character of utopia,[15] and interpreted Thomas More's essay as the manifestation of the possibility of a massive denial of all the locations on which the old *nomos* of the earth was based. He also interpreted the meaning of the word Utopia not simply as the non-place, *nowhere*, but also as the *u-topos* par excellence, a denial compared to which even the *a-topos* has a bond, albeit negative, with the *topos*. The second one is Françoise Choay, who recognized in Utopia the same root of abstract and calculating representation that characterizes modernity.[16] Utopia is neither the neutralization of the individual concreteness of places and landscapes, stories and time, nor the transformation of the land in an undifferentiated space. Instead, it emerges both from the limit and from the openness of such individuality, precisely from the symbolic imprint that every place contains within its borders. Utopia tells us about the possible fulfillment toward which a place tends, that is the possibility of a place to transgress itself, to leave its own stasis in order to aim toward a good condition that still is not, an adaptation to a greater accomplishment. The Utopian consciousness lives on a truly symbolic imagination and does not claim to fill up the gap between determinacy and the whole horizon of meaning in which it stands.[17] For this reason it is authentic and it is different from the anti-utopias and the dystopias, as well as from ideology.

Therefore, if the symbolic dimension characterizes the meaning of places, the art of shaping the built environment, what we call architecture, does not only express solidity, usefulness, and beauty, but also the overall meanings of the sense of being in the world. These meanings even determine the composition, orientation, geometry, proportion, and decoration. They express themselves, for example, in the expansion of space starting from a center, in the organization according to a vertical axis that is combined with the two axes of the horizontal plane forming the cross of the six directions (the four cardinal points, zenith and nadir), and in the relationship between the circle (and sphere), the most dynamic and least differentiated geometric shape, and the square (and cube), the most static shape.[18] We build because we dwell, as Heidegger states, subverting the platitude according to which we usually think about the art of building and cultivating.[19]

The Question of Identity

Places are historical-geographical configurations. They express history, character, and long-standing structures. They also give life to physiognomic identities (types) and to territorial individualities, which convey what we defined as the second meaning, the symbolic dimension. For this reason, if we interpret spaces only from a purely functional point of view, we will not be able to understand them fully. What is also necessary, therefore, is a description of the identity that considers the natural and sociohistorical elements, the aesthetic-perceptual dimension, the archives of knowledge, traditions, memory, and potential projects that are their heritage. The question is to grasp the texture between persistence and referrals, which hint, at first, at other times and other places, and ultimately point to a whole world of sense and fullness of being.

Dwelling in places is the way in which we stay in this constellation and we look after it. It is a practical relationship that also involves knowledge and affection, because places are texts that are constantly read, written, rewritten, and interpreted. The "consciousness of place" is the dizziness deriving from that concrete objectivity that cannot be reduced to objectification. It has been defined as the awareness—acquired through a process of cultural transformation of the inhabitants—of the territorial heritage, that is to say, the value of common goods (material and relational) as essential elements for the reproduction of individual and collective, biological and cultural life. In this awareness, the path from individual to collective connotes the elements that characterize the reconstruction of communitarian elements as open, relational, and supportive forms.[20] The reappropriation of the consciousness of place is therefore understood as the necessary condition for individual and collective construction in the forms of territoriality and self-sustainability.

The question of identity arises when we imagine an established community and a place with a specific, physical, economic, and cultural connotation. Even though we are considering it here in a nonclosed and nontribal way, there are also very different conceptualizations of the relationship between a place and its inhabitants. In fact, a place is often read—as though it were self-evident—as a safe haven, as a partly consistent space from the cultural point of view, as a detached area. This reading has the downside of interpreting the contemporary processes of transformation as disarticulations, as the rise of a "space of flux" that replaces the meaning of "space of places."[21]

However, places cannot be interpreted as coherent, stable, and detached realities. And this was true in the past as much as today. When it comes to the identity of places, they can only be interpreted as the result of a long history of relationships with other places. Therefore, they must be read as particular interrelationships in a broader field, and thus may leave open the possibility both to appreciate their local originality and to recognize wider ties.[22] They must be interpreted as essentially open and permeable, as the product of ties and connections, rather than closed, separate, and exclusive elements. In these intersections place acquires both its uniqueness and its interdependence on other places. The same can be said for human beings. Personal and group identities cannot be understood according to an essentialist-naturalistic paradigm, because they are neither an immutable heritage, nor a simple and accomplished reality, with no deficiencies.

If the search for purified identities produces a geography of rejection, the idea of an open and interactive identity means the possibility of creating a geography of reception,[23] which corresponds more truly to the reality and to the life of places and human beings.

On Ethics and Politics

Understanding the human condition as transcendentally allocated and interpreting place as a symbolic reality and a nexus of intersections in a functional, aesthetic, affective, cognitive, and meaningful relationship with those who inhabit it brings about some significant implications for ethical and political reflection.

First of all, considering localization as the constitutive mode of human existence can offer an important perspective for ethical thinking in order to understand the connections between the dimensions that we call particular, relative, objective, and common. Moreover, in spatial experience we can interpret the essential relationships between self and other, familiar and alien, between what is private and what is inappropriable. This confirms what we can call "human amphiboly." Being localized means being divided from others: every human being exists in a "here," according to which others are "there." Relationships with others are defined by such separation and the familiarity always takes place at a distance. Being localized implies an essential relationship with our "here," that is the place "outside us" that hosts us.

Being localized means that we orientate ourselves in the world starting from "our world," according to a "subjective where" that is neither private nor one that we own together, something that is nonexcludable but rival, such as common goods. At first sight, the possibility of emancipating oneself from a "here" seems to be a positive feature. For example, it is believed that real knowledge can be achieved only if it is detached from a "here." This claim is intriguing for thought and speculation. But if we really take into consideration human spatiality, we find out that real emancipation from our localization cannot be attained by merely neglecting to acknowledge the fact that we are always located in a space. Moreover, it may be exactly the attention to locality that leads us towards a human correlation between the relative and the universal. The relationship between the relative and the universal is revealed in the concrete experience of the "here" within a horizon with which it does not coincide, in its correlation with many other "theres" that represent the "here" of other people, and in the concrete experience of boundaries that are subverted in the effort to establish them.

However, the movement of thought has to be retraced in the opposite direction as well: ethical reflection can affect the practices that have to do with space. It is interesting to investigate what ethical theory concerned with human localization has to say about dwelling and about the building of places for human life. As a matter of fact inhabited places are often constructed places. However, there are also constructed places that are not meant for dwelling, as sacred places demonstrate, and we can live in places that are not built. We have to consider the ways of dealing with the places of human experience, that is to say, to investigate what distinguishes not only place, building, territory, house, city, and sacred space but also countryside, desert, and mountain.

This investigation deals with ethics but also has a considerable relevance for collective life, be it intended as politics, which is related to the dynamics of acquisition of power, or as policy, related to laws that manage public affairs, or as polity, which is related to the dynamics of social cohesion. The transcendental allocation of human life brings about issues related to place not only in relation to a communitarian approach, but also if we consider the practice of public life as the art of *being together among strangers*.

Here it becomes relevant that places are always interpreted by particular positions and reasons, and that the sense of place is conveyed not only by different media, such as novels, films, advertisements, paintings, but also by landscapes and by the decisions of domestic and foreign policy. A place can also refer to multiple geographic scales and involve different groups of people in different ways, that is to say, the same place can acquire different meanings for different people. These different meanings of place involve not only a consideration of past history, but also a vision of the future, and they may contradict each other and become sources of conflict.

We must not ignore the fact that interpretations are also carriers of social relationships, in a way that cannot be predetermined and that not only reflect but also affect relationships of power, which are sometimes latent. We can neither refer to the identity of place as a support for any form of conservatism (in the style of *Blut-und-Boden*), nor interpret it as an a priori factor to recognize the positive value of difference. It is very difficult to claim that the local inhabitants have some sort of exclusive right to control or even interpret certain parts of the earth, simply because it is not so easy to define who the local inhabitants are. And this is precisely one of the issues at stake. It may occur that those who "came before" are most ready to understand local value as mere resources, tools of wealth, without taking into consideration the risks to human health, landscape, soil fertility, and social cohesion.

Today the question of the meaning of places is no less important than in previous eras, due to the greater and faster displacements of individuals and populations. In fact, the common belief that states that at the time of globalization places are no longer important in their peculiarities does not seem true. Rather, these places are "put to work" on the basis of what constitutes their specific heritage or weakness in the present structure of unequal development. The uniqueness of places may also mean interconnected inequalities.

The question is to highlight, from time to time, the geography of power and the specific connection between social and spatial space as inescapable factors of the strategic scenario according to which we want to build our future.

Notes

*This original contribution has been translated into English by Elisa Piovesana.

1. E. S. Casey, *The Fate of Place* (Berkeley: University of California Press, 1997); J. E. Malpas, *Place and Experience* (Cambridge: Cambridge University Press, 1999).
2. V. Melchiorre, *Corpo e persona* (Milano: Vita e Pensiero, 1970), 53–92.
3. E. Levinas, *Totalité et infini* (Nijhoff: La Haye, 1971), 125–49.
4. M. Merleau Ponty, *Phénoménologie de la perception* (Paris: Gallimard, 1945), 294.
5. E. Cassirer, *An Essay on Man* (New Haven: Yale University Press, 1944).
6. M. Heidegger, "Bauen, Wohnen, Denken," in *Vorträge und Aufsätze* (Pfullingen: Neske, 1954), 145–62.
7. J.-Y Lacoste, *Expérience et absolu. Questions disputées sur l'humanité de l'homme* (Paris: Presses Universitaires de France, 1994), 10.
8. Levinas, *Totalité et infini.*
9. Lacoste, *Expérience et absolu,* 42.
10. Lacoste, *Expérience et absolu.*
11. P. Ricoeur, *La métaphore vive* (Paris: Editions du Seuil, 1975), 289.

12. M. Eliade, *Das Heilige und das Profane: Vom Wesen des Religiosen* (Hamburg: Rowohlt, 1957); V. W. Turner, *The Ritual Process: Structure and Anti-Structure* (London: Routledge & Kegan Paul, 1969).
13. R. Tagliaferri, *Saggi di architettura e di iconografia dello spazio sacro* (Padova: Edizioni Messaggero, 2011), 86.
14. R. Guardini, *Vom Geist der Liturgie* (Freiburg i.B: Herdersche Verlagshandlung, 1918).
15. C. Schmitt, *Der Nomos der Erde: Im Völkerrecht des Jus Publicum Europaeum* (Köln: Greven, 1950), 149–50.
16. F. Choay, *L'urbanisme, utopies et réalités, Une anthologie* (Paris: Editions du Seuil, 1965), 36–46.
17. V. Melchiorre, *La coscienza utopica* (Milano: Vita e Pensiero, 1970), 88–143.
18. A. Snodgrass, *Architecture, Time and Eternity* (New Delhi: Aditya Prakashan, 1990).
19. Heidegger, *Bauen, Wohnen, Denken.*
20. A. Magnaghi, *Il progetto locale. Verso la coscienza di luogo* (Torino: Bollati Boringhieri, 2001), 58–75.
21. M. Castells, *The Informational City* (Basil: Blackwell, Oxford 1989).
22. D. Massey and P. Jess, eds., *A Place in the World? Places, Culture and Globalisation* (Oxford: Oxford University Press, 1995), 221, 231.
23. I. M. Young, "The Ideal of Community and the Politics of Difference," in L. Nicholson, ed., *Feminism/Post-Modernism* (London: Routledge, 1990), 300–23.

Section II

Representing Space in Ancient and Medieval Contexts

Chapter 5

Analyzing Embodied Space in Ancient Israel's Tabernacle: A Biblical Studies Perspective

Mark K. George

Temple building was a common activity in the ancient Near East, something undertaken by kings on earth and deities in the heavens. In both Israel and Mesopotamia, temple-building activities were linked to the act of creation of the cosmos. In the *Enūma Eliš*, from Mesopotamia, creation is the result of a cosmogonic battle between two gods, Marduk, the victor, and Tiamat, the loser. At the end of this battle, Marduk creates the cosmos, and when he finishes, a group of lesser gods build him an earthly temple in Babylon in which to dwell. Building this temple is the culminating act of creation. In Israel, one of its creation stories (Gen. 1:1–2:4a) provides an account of the creation, yet no temple or earthly dwelling is built for the creator God. That part of creation is delayed until the latter half of the book of Exodus (chs. 25–31; 35–40), when Israel receives instructions for construction of the wilderness tabernacle. Once the people finish building it, the deity enters it and dwells among them (Exod. 40:34).

The mythical, symbolic understanding that temple building is linked to creation is not the only connection between temples and the cosmos for Israel. Jon Levenson argues that divine dwellings, whether temples or Israel's tabernacle, are depictions of the world, in the sense that they are a microcosm (literally, a micro-cosmos) of the world and cosmos. Conversely, the cosmos is a macro-temple (or, -tabernacle).[1] The social and cultural meanings of temples thus involve a scale far beyond that of these buildings' physical spaces, mental arrangement (e.g., the floor plan, elevation, location in a city or town), and other such realities.

While the symbolic connections between temple or tabernacle and creation have been recognized, the relation of the human body to them has tended to be overlooked. There is a direct connection between human beings, particularly the human body, and the symbolic meanings of temples. Temples literally are built on a human scale, because measurements and dimensions in Israel and elsewhere in the ancient Near East were determined by reference to the body, such as hands and fingers. Thus, to the extent temples are linked to the cosmos, so too is the human body. If the tabernacle is a microcosmos, then, that cosmos is scaled to the human body, or at least is capable of being understood in relation to it. For the Priestly writers, whom scholars widely agree wrote the Genesis 1 creation account and the tabernacle narratives, the human-cosmos connection apparently is as it should be, since human beings are made in the image of God at the end of creation (Gen. 1:26–27).

The connection between the tabernacle's symbolic meaning and the human body has implications for the interpretation of the narratives about them. One is that the human body in tabernacle space merits scholarly attention and analysis, or at least more than it previously has received. Another is that such analysis must be attentive to how the tabernacle's symbolic meanings relate back to the body. Put another way, the meaning of tabernacle space is inseparable from the human body. Understanding how this works is something that spatial theory, particularly theories of embodied space (such as those of Michel de Certeau and Henri Lefebvre), can help explain. This theoretical perspective draws attention not only to how the body is of practical use in conceptualizing the tabernacle (e.g., for measuring), but also how it is assumed in, and might experience, that space. The body perceives tabernacle physical space via its senses, it is the standard by which it is planned and organized, and it is the site where its significations and social meanings are lived and experienced. Attention to the body enables greater appreciation of tabernacle space as a religious symbol system. That symbolic system, connecting bodies and the cosmos, served as a social map of the world, one filled with meaning.[2] For the Priestly writers, this means an individual Israelite who understands her or his social location in the tabernacle simultaneously can understand her or his place within the cosmos. Such a possibility might be important in any number of social and historical circumstances, but especially so in a time of exile and displacement from one's country, as it was for the Priestly writers.[3]

In what follows, I examine tabernacle space from an embodied perspective, in which the presence of the human body in this space is my primary concern. I begin with a brief summary of the narratives before proceeding with my analysis.

Israel's Wilderness Tabernacle and Its Ritual Importance

Ancient Israel's tabernacle is described in thirteen chapters of Exodus, almost one third of the book's length. The tabernacle is portrayed as a portable complex consisting of a tent structure and surrounding open-air court. The tent structure, or tabernacle proper, is made from four coverings (of linen with cherubim worked into them, goats' hair, tanned rams' skin, and another, uncertain, material) set on wooden frames placed in metal bases with bars between the frames for support (Exod. 26; 36). A curtain within this structure (Exod. 26:33) divides it into two: most holy space and holy space. These spaces contain several specific items. Most holy space, on the western end, contains the ark and *kappōret* (mercy seat; Exod. 26:34; 40:21). Holy space, on the other side of the curtain, contains a lampstand, the table for the Bread of the Presence, and an incense altar (Exod. 26:35; 30:6; 40:22, 24, 26).

Surrounding the tabernacle proper is an open-air court created by a set of linen hangings on wooden frames set in metal bases with bars between the frames for support (Exod. 27; 38). In the court are the altar of burnt offerings, the bronze basin, and the tabernacle proper (Exod. 30:17; 40:29–30). The narratives state that the tabernacle proper is located inside the court, as are the altar and basin, but their exact placements are not given, which is surprising for ritual space, since such placements presumably are important. The basin is closer to the tabernacle proper than is the altar (Exod. 30:18), and both stand between the tabernacle proper and the court entrance, but no further details of their locations are provided. As a result, modern biblical scholars tend to favor a geometric arrangement on the central east-west and north-south axes, with the tabernacle proper, basin, and altar centered on the east-west axis. But such geometric precision is not specified by the texts. Indeed, there is nothing to suggest it is impermissible to place the tabernacle proper in the northwest or southwest corner of the court, with the bronze basin somewhere between (not necessarily directly in front of) its entrance and the court entrance, and the altar between the basin and court entrance.[4] For example, the tabernacle proper could be in the southwest corner of the court, the bronze basin in front of it somewhere along the south wall, and the altar along the north wall.

Within the tabernacle proper, the placement of objects is somewhat more precise. The ark and *kappōret* are placed behind the curtain in most holy space, but their exact locations are not specified (against the back wall? in the middle? in a corner? up against the curtain?). This is surprising, since these two objects are most closely associated with the deity, and the narratives do concern themselves with other ritual details in this space, such as how to dispose of sacrificial blood (e.g., Exod. 29:12). In holy space, the incense altar is located in front of the curtain (Exod 30:6), the lampstand is on the south wall, and the table on the north (Exod. 26:35). Other than these general descriptions, however, the placement of these items is left to the imagination. Once again this is surprising for ritual space.

Tabernacle space does not end at the court walls. Its conceptual logic classifies and organizes space outside the court into three distinct zones. In the first zone, the Levitical clans are arranged immediately outside the court walls along its north, west, and south sides (Moses, Aaron, and their families are positioned on the east side, in front of the court entrance; Num 3). In the second zone, beyond the Levitical tribes, is "camp space," in which are arranged the twelve tribes of Israel in a particular order (Num 2). The third zone is space "outside the camp," beyond the twelve tribes, where Israelites in a state of ritual uncleanness or impurity dwell temporarily, where the offal from sacrifices is taken, and where the rest of the peoples of the world dwell.[5]

Scaled to the Body

The tabernacle narratives are highly conceptual and prescriptive; they are mental space. Scholars generally accept that conceptual perspective, thereby leaving the question of a person or body in space as a marginal concern at best, and rendering the tabernacle an empty, impersonal space.[6] Yet there are numerous ways in which the presence of the body in tabernacle space is assumed in the narratives.

As noted above, the measuring system for distances is referenced to the human body: cubits, spans, handbreadths/palms, and fingers. A cubit is the distance from a man's elbow to the tip of his index finger, a span the distance of the open hand from thumb to fingers, a handbreadth/palm the width of four fingers, and a finger the width of one finger.[7] Because dimensions are central to how tabernacle space is described, so too is the body, upon which those dimensions are scaled. The ark, for example, is 2.5 cubits long, 1.5 cubits wide, and 1.5 cubits high (Exod. 25:10–11; 37:1–2). The table for the Bread of the Presence is 2 cubits long, 1 cubit wide, 1.5 cubits high, and has a rim "a handbreadth wide" around it (Exod. 25:23–25; 37:10–12). And so on. Each of the objects crucial to the creation of tabernacle space, from the ark to the hangings of the court, is described using dimensions referenced to the human body. They have a human scale.[8]

Various bodily senses are important to tabernacle space. The tabernacle has a particular visual appearance. The materials used to create tabernacle space (e.g., metals, textiles, fabrics, wood), their ornamentation (e.g., cherubim worked into certain linens, moldings and rims, flower petals), how items are assembled, and other details imply visual perceptions of it. The ark, for example, is a pure gold covered acacia wood box, with a gold molding around the outside, four gold rings on its four feet (two on each side), and long acacia wood poles, overlaid with gold, that fit through the four rings on its feet (Exod. 25:12–14; 37:2–5). This list of perceptible details—dimensions, materials, construction, ornamentation, assembly—creates an image of the ark, even if they lack certain details, such as the thickness of the ark's walls or which direction the rings and poles face, resulting in interpretive ambiguity. The ark would be a visually recognizable object to a person seeing it in space, all the more so because of the *kappōret* that sits upon it. Fashioned of pure gold, 2.5 cubits long by 1.5 cubits wide, it has two cherubim on its top, one at each end, with their wings spread out over the mercy seat, made so that the whole is one piece (Exod. 25:17–20; 37:6–9). It is visually distinctive, even if its precise construction is uncertain. It would be a spectacular sight, with its gleaming gold and distinctive decorations, something even Steven Spielberg recognized in the *Raiders of the Lost Ark* movie.

Material details are not the only means by which tabernacle space is perceived. Lists within the narratives help focus and reinforce visual perceptions. Three types of lists serve this purpose. The first are lists summarizing the materials required to create the particular objects of the tabernacle (Exod. 25:1–7; 35:4–9). These lists serve what Umberto Eco calls a practical purpose, because they are limited and refer to real things outside the lists (things known to exist by the author).[9] The practical nature of the materials lists validates the objects' visual details. A second type of list summarizes the objects that constitute tabernacle space (Exod. 28:4; 31:7–11; 35:10–19; 39:33–41). They refer to the particular, limited set of items of the tabernacle: ark, *kappōret*, four coverings, frames, bars, bases, curtain of the tabernacle proper, table, incense altar, lampstand, altar of burnt offerings, bronze basin and stand, hangings, frames, bases, bars of the court, priestly vestments, oil, and incense. These items are described in terms of dimensions, materials, and decorative features and ornamentations.[10] In addition to being practical, they are what Eco terms "pragmatic," because they constitute a checklist of the tabernacle.[11] Tabernacle space *is* all of these things, and each of them must be brought together for it to constitute tabernacle space. The absence of any of these items changes the space; it is "almost, but not quite" the tabernacle. Conversely, no single item or incomplete grouping of items is "the tabernacle." The ark alone is not "the tabernacle." Neither is the assembly of all these objects *without* the ark "the tabernacle."[12] Visual confirmation of the presence of all these items would enable a person to know that he or she is in Israel's tabernacle space. The third type of list details the skills required to create the tabernacle (Exod. 31:1–6; 35:30–35). They reinforce visual perceptions of tabernacle space by confirming and explaining the skill and handwork abilities required to create the items constituting the tabernacle.[13]

Vision plays a role in perceiving how tabernacle space is configured. Approaching the tabernacle complex from the east, from the space "outside the camp," an individual would experience the arrangement of tribes around the complex long before reaching, perhaps even seeing, it. Upon arrival, the challenge of entering the court would be an immediate problem. There are designated entrances/exits (curtains and screens; Exod. 26:31–37; 27:16; 36:35–38; 38:18–19) into the court, tabernacle proper, and most holy space. These are the mechanisms used by everybody to move through the physical demarcations of space (i.e., its walls) in the complex. How a body does so—going around or under them—is not explained. Once inside court space, the placement of the tabernacle proper, bronze basin, and altar of burnt offering would become apparent, as these items are found in court space. So too would the lack of furnishings, which suggests the court is meant to have as many people standing within it as possible. No designated routes are marked out in this space. Standing in the court, a person could look out, above the court hangings, but not immediately next to them; no elevated positions are described within the court. By contrast, activities in the tabernacle proper are shielded from view to anyone outside it, due to its coverings. Only priests may view what occurs within it and its configuration: lampstand on the south wall (the left as one enters), table on the north (the right), and incense altar in front of the curtain.[14] Only the high priest, on the Day of Atonement (Lev. 16), could enter most holy space and see the placement of the ark and *kappōret* behind the curtain. One thing that *would* be apparent to people in the court is that only some people are able to enter the tabernacle proper. Their different spatial access is marked physically by their attire: tunics, sashes, headdresses, for the priests; an ephod, breastpiece, robe, checkered tunic, sash, and turban with a pure gold rosette for the high priest. Spatial configurations have corresponding social configurations, and clothing marks certain bodies for certain spaces.

Various locations within tabernacle space are centers of activity that focus attention and intensify space. The altar of burnt offering is the most intensively used space because the greatest amount of activity occurs there. Most sacrifices, offerings, and other rituals involve it in some way. The entrance of the tabernacle proper is another locus of activity and intensification of space. These two sites are where bodies (human and animal) congregate and concentrate, and ritual actions occur.[15] Holy space within the tabernacle proper also is a focus of attention: the priests tend the lampstand lamps every evening and morning (Exod. 27:20–21); the high priest offers incense on the incense altar morning and evening (Exod. 30:7–8), and bread for the table for the Bread of the Presence is placed on it "always" (Exod. 25:30). On the Day of Atonement each year, the high priest "makes atonement" on the incense altar, ark, and *kappōret*, daubing them with blood (Exod. 30:10; Lev. 16:14–15).

Finally, vision is important for recognizing two other features of tabernacle space. Cardinal directions determine how the tabernacle complex is set up and practiced. Cosmological factors—the Sun and Moon—would assist in determining directions, something made easier by the open court. All the entrances/exits face east. Most holy space is at the western end of the complex. By determining the location of objects in terms of cardinal directions, tabernacle space explicitly links a person's experience of it with the cosmos and cosmological symbolisms. Cardinal directions are important because the tabernacle is portable. Everything within it is designed to be taken apart, packed up, moved, and set up again at a new site.[16] This renders some items functionally impractical, such as the altar of burnt offering with its wooden core (acacia wood overlaid with bronze; Exod. 27:1–2; 38:1–2). Biblical scholars long have noted that the core would disintegrate if a fire were set in the altar.[17] But this impracticality emphasizes the importance of portability to tabernacle space, which is designed to be moved and recreated, using the cardinal directions as a guide for how to reconstitute it. It also indicates the people who use this space are capable of moving; they are not a settled people.

Other bodily senses also operate in tabernacle space. Touch is involved when people lay their hands on sacrificial animals (e.g., Exod. 29:10, 15, 19; Lev. 3:2, 8, 13), help clothe the high priest Aaron and his sons during their ordination ceremonies (Exod. 29:4–9), handle the blood, flesh, viscera, and offal of the sacrificial animals (something the priests do; e.g., Exod. 29:12–14; Lev. 1:5–6, 8–9), and arrange wood and fire on the altar of burnt offering (e.g., Lev. 1:7). The smoke of sacrifices is "a pleasing odor to the LORD" (e.g., Exod. 29:18, 25, 41; Lev. 1:9, 13, 17), the incense of the incense altar is "fragrant" (Exod.

30:7), and certain ingredients (e.g., cinnamon and aromatic cane, Exod. 30:23) of the incense and anointing oil explicitly are named, presumably because of their smell. Other spices of the incense and anointing oil, such as myrrh, cassia, olive oil, and frankincense (Exod. 30:24), likely excited the sense of smell as well. Other, unnamed smells are inferred: both fresh and rancid blood, animal offal and viscera, animal excrement, sweat from animals and humans, and burnt flesh from the sacrifices. Because the court was not covered, presumably there would be an exchange of smells from tabernacle space, the camp, and beyond. Tastes occasionally were present: certain offerings and sacrifices could be eaten by the priests (e.g., Lev. 6:16; 7:15–16), and food regularly was consumed in camp space. The one sense that is relatively rare is sound. Bells on the high priest's robe were heard annually when he went into most holy space, to signal he was alive (Exod. 28:35). Moses announced to the congregation various instructions, including how to use the anointing oil (Exod. 30:30–33), what they were to construct (Exod. 35:1, 4), who would lead the construction (Bezalel and Oholiab; Exod. 35:30–36:1), and the introductory words during the ordination of Aaron and his sons (Lev. 8:5). Leviticus is filled with Moses's words to Israel about sacrifices, offerings, and other matters of cultic life, which he speaks in front of the tabernacle proper. Yet no words are recorded as being spoken over sacrifices and offerings in tabernacle space, as others have noted.[18]

Body and Cosmos

The human body is a central assumption in the conception and creation of tabernacle space. From how it is measured, to the ways in which it is experienced, the body is integral to it. Much of the way tabernacle space is described appears to assume a priest's body and perspective, because of details only a priest could know. Access to the court is permitted for both male and female members of the Israelite congregation, that is, all those who serve (i.e., worship) Israel's deity through observance of sacrifices, festivals, and other obligations.[19] Thus any member of the congregation could perceive and experience the range of activities, people, objects, and placement of items in court space. The placement of the lampstand, table, and incense altar in the tabernacle proper, however, is something only a priest could know. Likewise, only the priests (presumably) would know about the linen undergarments worn by Aaron and his sons as part of their priestly attire.[20] It seems likely only the priests heard the sound of the bells on the high priest's robe, since they may not have been audible outside the four coverings of the tabernacle proper, whereas they would be through the curtain inside it. The lack of details about the placement of the ark and *kappōret* in most holy space argues against the narratives being from the perspective of the high priest.

The narratives' priestly perspective is representative of the social nature of tabernacle space and the social preferences and distinctions encoded into it. The multiple spatial zones that constitute tabernacle space—outside the camp, camp, Levites, court, holy, and most holy—are socially constructed and operate on the basis of certain social distinctions. Only certain bodies may be in particular spaces. Noteworthy is that the boundaries between the tribes and Levites in camp space are not sharply defined, suggesting people from other tribes can transgress those boundaries by crossing them, even if they may not *camp* within different boundaries. Neither does a sharp boundary exist between camp space and space outside the camp, suggesting foreigners (nonmembers of the congregation) may move through camp space. The imprecision of boundaries symbolizes Israel's fluid, porous relationship with other peoples of the world and each of the tribes. Distinctions in the camp, tabernacle, and thus creation and the cosmos, are present, but porous. There is a system, but it is not rigid!

Boundary imprecision, however, is not found within the tabernacle complex. Entry into court space is restricted to members of the Israelite congregation. Even more restrictive is entry into the tabernacle proper, and most restrictive is entry into its most holy space. Social factors, which identify bodies in particular ways, serve as the taxonomic criteria operating at each boundary determining which bodies may enter a space. As a micro-cosmos, the system dividing and categorizing tabernacle space suggests that social factors determine divisions among humans within the cosmos.[21] Finally, it is noteworthy that there are no steps, ramps, or other features of the tabernacle complex that enable one person symbolically to be elevated above another. The court is not higher than the camp, nor the tabernacle proper from the court, nor most holy space from holy space. Barry Schwartz argues that verticality often signals social, hierarchical distinctions.[22] The cosmos re-presented by the tabernacle organizes such distinctions differently. Tabernacle space flips vertical classifications on their side, encoding social distinctions horizontally, by access to spatial zones. To the extent the tabernacle is a micro-cosmos, that cosmos is understood and lived on a horizontal plane for the Priestly writers. Of course, with the deity symbolically inhabiting the tabernacle's most holy space, creator and creature were both dwelling on the same plane, suggesting a form of commonality, even if not equality (marked by the various curtains and screens at the boundaries).

Conclusion

The human body is an important element of tabernacle space, and merits greater scholarly attention. The body is both explicitly and implicitly present in the tabernacle narratives, whether through the ways in which this space is measured and visualized, or the use of touch and smell. It also is present in the conceptualization and organization of this space. Bodies are marked, classified, and organized in ways that are replicated and realized in distinctions between tabernacle spaces, both inside and outside court walls. Who can enter or camp in a particular space is determined by socially determined characteristics of a body, such as tribal affiliation, gender, or family lineage. A linkage therefore is created between bodies and tabernacle space, with the social characteristics that are mapped onto bodies simultaneously serving as a map of the tabernacle. Given that the tabernacle is a microcosmos, mapping the body onto tabernacle space serves as a means of mapping the body onto the cosmos. These meanings and associations all are part of the religious symbol system that is tabernacle space.

"Religion is an inextricably human phenomenon," argues Jonathan Z. Smith; it is a means of constructing reality and meaning in the world.[23] Religions and religious symbol systems therefore are part of the human quest to understand and interpret the world within which a person lives, so as to give life significance.[24] The social space of Israel's wilderness tabernacle is one means of creating an orderly, consistent world and cosmos for the Priestly writers, even while they find themselves living in exile, dis-placed. It is within this understanding of tabernacle space as a religious symbol system that the tabernacle's importance for Israel's cult and rituals, so often cited as the reason for these detailed narratives by biblical scholars, must be understood.[25] Performance of Israel's cult through its rituals simultaneously creates and reinforces tabernacle space in all its social dimensions. The Israelite body, assumed in tabernacle space as the reference point upon which it is created, constructs a world of meaning for Israel, one in which the Israelites can dwell wherever they find themselves.

Notes

1. Jon D. Levenson, *Creation and the Persistence of Evil: The Jewish Drama of Divine Omnipotence* (San Francisco: Harper & Row, 1988), 86. Cf. also Carol L. Meyers, *Exodus* (New York: Cambridge University Press, 2005), 224.
2. On religions as meaning-making, cf. Jonathan Z. Smith, "Map Is Not Territory," in *Map Is Not Territory: Studies in the History of Religions* (Chicago: University of Chicago Press, 1978), 289–309.

3. I assume an exilic date for these writers.
4. Cf. the proposal of William H. C. Propp, *Exodus 19–40: A New Translation with Introduction and Commentary*, AB 2A, David Noel Freedman, ed. (New York: Doubleday, 2006), 498.
5. I explain in detail this arrangement in my book, Mark K. George, *Israel's Tabernacle as Social Space,* SBLAIL 2, Benjamin D. Sommer, ed. (Atlanta: SBL Press, 2009), Ch. 4.
6. E.g., John I. Durham, *Exodus* (WBC; Waco, Tex.: Word, 1987), 350–53; Terence E. Fretheim, *Exodus* (Louisville, KY: John Knox Press, 1991), 263–68, 272–74; Meyers, *Exodus*, 219, 222–23; Propp, *Exodus 19–40*, 687–89.
7. Marvin A. Powell, "Weights and Measures," *ABD* 6:897–908. Given physical variation between individuals, distance itself is variable in tabernacle space; no standard for these measurements is stated in the texts.
8. By comparison, the serving utensils, such as forks and plates, are not described in detail, whether number, materials, or size.
9. Umberto Eco, *The Infinity of Lists* (New York: Rizzoli, 2009), 113–29. Because the tabernacle may not have existed physically in history, Eco's statement must be modified here to include potentially real objects. For a review of the scholarly debate concerning the tabernacle's historical existence, see George, *Israel's Tabernacle*, 11–13.
10. The exception is the bronze basin, about which only the materials and pieces (a basin and stand) are detailed, along with the enigmatic notice that the bronze comes "from the mirrors of the women who served at the entrance to the tent of meeting;" Exod. 30:18; 38:8.
11. Eco, *The Infinity of Lists*, 113–29.
12. The pragmatic nature of these lists provides order to tabernacle space, and is another symbolic means that links tabernacle space and the cosmos (cf. Levenson, *Creation and the Persistence of Evil*, 86). The predictable and orderly nature of the tabernacle, which consists of an identifiable set of items, makes the symbolic claim for predictability and orderliness in the cosmos.
13. If the tabernacle never existed in history, then the lists may be considered "poetic," because they conform to the authors' own constraints. These constraints likely were literary, and thus formal, because they echo lists of materials, labor, and items for temples found in temple-building narratives elsewhere in the ancient Near East. Cf. George, *Israel's Tabernacle*; Victor (Avigdor) Hurowitz, *I Have Built You an Exalted House: Temple Building in the Bible in Light of Mesopotamian and Northwest Semitic Writings*, JSOTSup 115 (Sheffield: JSOT Press, 1992). Lists of items not specifically detailed individually in the narratives suggest the importance of these lists for focusing visual perceptions. Plates and dishes for incense, flagons and bowls for drink offerings (Exod. 25:29; 37:16), pots for ashes, shovels, basins, forks, and fire pans for the altar of burnt offering (Exod. 27:3; 38:3), are

mentioned as part of the tabernacle, but not further described. They are seen *in*, but not *as*, the tabernacle.

14. Notably, the Hebrew expression for "in front of" is referenced to the body. Literally, the term is "to the face of," *lipnê*.
15. Leviticus locates the performance of numerous rituals at these sites.
16. Moving the tabernacle and camp involves a particular sequence, with the tabernacle items located in the middle of the marching people; Num 2:17.
17. Durham, *Exodus*, 357–76; Meyers, *Exodus*, 237; Nahum M. Sarna, *Exodus* (Jewish Publication Torah Commentary; Philadelphia: Jewish Publication Society, 1991), 172.
18. Levenson, *Creation and the Persistence of Evil*, 95; Propp, *Exodus 19–40*, 497.
19. Access for men and women in the court is a debated point among scholars, and not all biblical texts are in agreement about whom may enter tabernacle or temple spaces such as the court. See my discussion of the composition of the congregation in George, *Israel's Tabernacle*, 112–19.
20. It is conceivable a priest's wife or children would know this, too, although it is possible these garments remained in tabernacle space; cf. Lev 6:11 [6:4 MT] and the arguments of Jacob Milgrom, *Leviticus 1–16: A New Translation with Introduction and Commentary*, AB 3 (New York: Doubleday, 1991), 386.
21. On dividing and separating being part of the design of the cosmos, see Frank H. Gorman, Jr., *The Ideology of Ritual: Space, Time and Status in the Priestly Theology*, JSOTSup 91 (Sheffield: JSOT Press, 1990).
22. Barry Schwartz, *Vertical Classification: A Study in Structuralism and the Sociology of Knowledge* (Chicago: University of Chicago Press, 1981).
23. Smith, "Map Is Not Territory," 290.
24. Ibid., 290–91.
25. Biblical scholars generally interpret the numerous details about the tabernacle as resulting from a priestly concern with both the cult and its rituals.

Chapter 6

Sacred Spaces in the Book of Hosea: The Intersection of Gender, Space, and the Sacred in Ancient Israel from a Feminist-Theological Perspective

Christl M. Maier

Like no other prophetic book of the Hebrew Bible, Hosea presents a conflict about "true" ritual and "right" sacred spaces. The prophetic voice polemically criticizes local sanctuaries and characterizes the worshippers at these sites with pejorative feminized terms. The addressees of the prophetic critique, however, are not granted a voice so that the text also reveals a hierarchy of power, in which the prophet can define what is right or wrong. Both the intention of Hosea's polemic and the rituals performed at these sites are highly disputed among scholars. My analysis of space and gender in some passages of the book aims at highlighting the sociohistorical reality of these sacred spaces mirrored in Hosea, where competing religious convictions intersect. Rather than taking a stance in the conflict I will describe the different perspectives and critically analyze their underlying ideologies. The multidisciplined character of this volume requires me briefly to introduce my methodology with regard to space and gender. The main part of this article analyzes passages of Hosea that deal with the local sanctuaries and the royal sanctuary at Bethel.

Spatial Analysis and Gender Analysis

In this study, I use the spatial theory of the French Marxist sociologist Henri Lefebvre (1901–1991) as a heuristic tool to analyze sacred space. Lefebvre's theory of space, published in French in 1974 and translated into English in 1991, has greatly influenced current geography, architecture, and sociology.[1] Lefebvre's epistemology of space focuses on the interrelatedness of spatial practice, conceptualization, and human experience of space. He argues that space is socially produced and "'incorporates' social actions, the actions of subjects both individual and collective."[2] His theory analyzes the production of space from three dimensions that are intrinsically intertwined: First, Lefebvre holds that space is produced by spatial practice (*pratique spatiale*), for example, by architecture and urban planning as well as by daily life and routine that encompasses both production and reproduction.[3] From the viewpoint of an individual, this spatial practice presupposes the use of the body, that is, hands, sensory organs, and gestures, and thus this realm of space is *perceived* as physical space, the materiality of space (*l'espace perçu*).[4]

The second dimension is named "representations of space" or "conceptualized space"[5]; Lefebvre here refers to the results of social processes of planning, naming, and inscribing of significance, thus, to space as mentally constructed or ordered, a process that is often shaped by a certain political strategy in society and thus dominates the concepts and ideas about space. On the individual level, this is space as *conceived* space (*l'espace conçu*) by scientific knowledge of anatomy, physiology, and so on.[6]

The third dimension analyzes the social significance of a society's space by focusing on *spaces of representation*,[7] which embody complex symbolisms, communal values, traditions, metaphors, dreams, that is, a different, often nonverbal collective experience. This is space as experienced or *lived* (*l'espace vécu*), "*lived* through its associated images and symbols, and hence the space of 'inhabitants' and 'users.'"[8] Lefebvre employs the plural pointing out that various simultaneous representations may exist as different groups or individuals appropriate space in various ways through imagination and symbolic use. This third dimension, space as experienced or lived, is important since it covers any use of space that challenges the dominant conceptualized space and thus counteracts hegemonic spatial patterns.

While one has to keep in mind that Lefebvre analyzed modern societies dominated by a capitalist mode of production in the 1960s, I find his theory useful to also analyze the production of space in an ancient society and culture, even if common knowledge of such society is mainly mediated through texts. I further argue that space as described in biblical texts comprises all three dimensions of space: such narrated space is produced by spatial practice that makes use of its materiality as well as by a certain ideology and experience of living in it.

What can be said of any space in text and reality also applies to sacred space. The term "sacred" goes back to the Latin word *sacer* (dedicated to a divine being), the antonym of which is *profanum* (outside of the sacred precinct, mundane, common).[9] As the historian of religion Jonathan Z. Smith has demonstrated, space is sanctified or made holy by humans through the cultural labor of ritual, in specific historical situations, involving the work of attention, memory, design, construction, and control of place.[10] Thus, sanctuaries as sacred spaces are the products of a specific religion in a particular cultural and historical situation. In my spatial analysis, I will demonstrate how a certain spatial practice produced a sacred space, the religious significance of which was contested in a time of crisis.

With regard to "representations of space," the ideology of space, gender analysis is vital, since the book of Hosea relates certain spaces to female bodies and uses female personifications for the people, the land, or the city. [11] The feminist theorizing of the body has demonstrated that ideas about body and gender are, like ideologies or symbolisms of space, closely tied to cultural and historical circumstances of a given society.[12] This corresponds to Lefebvre's insight that the human body as space is subject to the production of space in a given society.[13] Therefore, the use of the body in literary personifications and texts in Hosea is tied to societal norms and needs to be analyzed in relation to their sociohistorical context.

The Conflict about the Local Sanctuaries in the Book of Hosea

The most salient feature of the book of Hosea is its use of sexual metaphors for worship and politics.[14] The prophet Hosea and the subsequent authors of the texts that now form the book employ the Hebrew verbal roots *znh* (to prostitute oneself, to play the whore) and *n'f* (to commit adultery) as keywords[15] for their polemic dismissal of religious activities, the nature of which is highly disputed among scholars. Taken literally, prostitution and adultery are distinct concepts: *znh* (to prostitute oneself) refers to any sexual activity in exchange for payment. Although prostitutes are marginalized in Israelite society, prostitution is not illegal.[16] Adultery, however, is any sexual act of a betrothed or married woman outside of her marriage and considered a legal offense to the husband, which violates his property right and damages his reputation. Whereas in Chapters 4–14, the metaphor "such worship or politics is whoring" is used more frequently, Chapters 1–3 bring the concept of marriage and adultery to the foreground in a dramatic narrative about the prophet's marriage with Gomer, a former prostitute. Since Hosea 1–3 introduce the book, Hosea's marriage to Gomer represents the relationship between God and Israel and thus offers an interpretive lens to the collection of loosely linked prophetic announcements in Hosea 4–14.

Within the book, two different places of worship are mentioned although none is explicitly called "sacred." First, there is "the house of God" (Hosea 8:1; 9:4) at Bethel, which according to 1 Kgs 12:25–29 was built by the king of Israel to host a sacred object in the shape of a calf and thus is a national sanctuary.[17] Bethel literally means "house of (the god) El" but in Hosea is renamed Beth-Aven "house of disaster" in order to underline the prediction that the sanctuary will be destroyed (Hosea 4:15; 5:8; 10:5, 8). Yet, the primary conflict about sacred space revolves around sanctuaries called *bāmôt* (Hosea 4:11–14; 10:8), which are therefore central to my analysis.

The Local Shrines and Their Cults

Many scholars interpret Hosea 4 as a series of passages that present authentic prophecy[18] or as an authentic reflection of the prophet's time and situation.[19] Hosea 4:11–14 forms a unit framed by the "people" (v. 12, 14) and a double citation of proverbs (v. 11, 14).[20] As many scholars have noticed, the passage has a concentric structure as displayed in my translation:[21]

> *11* Whoredom, wine and new wine take away understanding.
> *12a* My people consults a piece of wood,
> and its staff gives oracles.
>> *b* For a spirit of whoredom has led (them) astray,
>> and they whored away from *beneath* their God.[22]
>> *13a* They sacrifice *upon the tops* of the mountains,
>> and burn incense *upon* the hills,
>> *beneath* oak, poplar, and terebinth because its shade is good.
>>> *b* Therefore your (masc. plural) daughters play the whore,
>>> and your daughters-in-law commit adultery.
>>> *14a* I will not punish your daughters when they play the whore,
>>> nor your daughters-in-law when they commit adultery;
>>
>> for they themselves go aside with the whores
>> and sacrifice with the *qĕdēšôt*;
>
> *b* thus a people without understanding comes to ruin.

The passage does not provide a prophetic oracle in its classical form but a polemical assessment of the people's behavior, which in v. 13b–14a abruptly turns into a direct address to a group of men. The term "my people" (v. 12a) and the announcement "I will not punish" (v. 14a) identify the speaker as the national deity of Israel. The people who seek advice from wood and staff (v. 12a) are viewed both as a collective body (with singular forms in v. 12a) and as a group of individuals (with plural forms in v. 12b).

Since my focus is the characterization and function of sacred space, the most important verses are 12b–13a, which offer a binary pair of locatives. The sacrifice takes place *upon* the mountain-tops and *upon* hills, but *beneath* huge trees. This cult is polemically dismissed as "whoring away from *beneath* their God" (v. 12b). This use of the book's key metaphor explicitly depreciates the addressees' cultic activities by sexualized language. Their rituals are judged as apostasy from the God to whom the people belong (v. 12a). Although the name of the national deity is not revealed, the passage indicts Israel's cult at the local sanctuaries as totally erroneous, without understanding (v. 11, 14), and the cause of their ruin. The negative assessment of the cultic activities also applies to the sacred sites, the production of which will be evaluated by using Lefebvre's three dimensions of space.

Spatial Practice or Perceived Space

Hosea 4:11–14 mentions a hillside open-air area with trees and some sort of wooden object (v. 12a) as well as rituals of sacrifice and burning incense (v. 13a), part of which are performed with the *qĕdēšôt* (v. 14a, sacred women). The verb "to sacrifice" (Hebrew *zbḥ*; v. 13a) refers to the ritual slaughtering of an animal. Animal sacrifice and the burning of incense require an altar as cultic installation. The notion of "consulting" and an object that "gives oracles" (v. 12a) refers to the practice of divination, namely to ask a deity for advice through media. The description of the rituals on the hillsides fits well to a well-known sacred site called *bāmâ* (plural *bāmôt*, cf. Hosea 10:8)—a local open-air sanctuary equipped with trees, an altar, a *maṣṣēbâ* (standing stone), and an *ʾăšērâ* (a wooden pole or trunk of a tree).[23] In his analysis of all biblical references to *bāmôt*, Matthias Gleis concludes that these sanctuaries are not necessarily located on hills, thus the common translation "high place" or "cultic height" is rather misleading.[24] They are situated in the vicinity of towns and are mostly open-air, sometimes with a small building that was probably used to store cultic vessels and to consume sacrificial meals (1Sam 9:22; Ezek 16:16). While the numerous *bāmôt* constituted an unchallenged part of the local religion both in Israel and Judah during the monarchy, they were later condemned as sources of idolatry.[25] Hosea 4:11–14 is probably the earliest critique of these sites, the sacredness of which is established by a certain ritual practice: the local population may gather in the shadow of a huge tree, sacrifice an animal on the open-air altar, collectively consume part of the sacrificial meat, and perform prayers or burn incense in front of the standing stone and the wooden pole.

For modern scholars, the most interesting question is what sort of ritual was performed and by whom. Rituals of divination through objects (v. 12a), animal sacrifice, and incense burning (v. 13a) are common features of almost any cult in the ancient Near East and thus rather unspecific. The sexual metaphors used in Hosea 4:11–14, however, led to the thesis that the prophet here condemns cultic prostitution.[26] The idea that cultic or sacred prostitution, that is, sexual intercourse performed within a sanctuary as a fertility rite, goes back to the classicist Sir James Frazer who in his work *The Golden Bough*, originally published in 1890, charged the ancient world with mystical sexual fervor and the desire to reproduce.[27] This Frazerian myth found wide recognition among Western scholars because it nurtured the nineteenth-century European prejudice of orgiastic cults and harems as being characteristic of the Near East.[28] In Hosea 4:14, the plural *qĕdēšôt* (singular *qĕdēšâ*, sacred woman) is paralleled to *zonôt* (whores) and therefore translated with "cult prostitutes" or "temple prostitutes" in many current English and German Bibles.[29] Many twentieth-century commentaries on Hosea either link the supposedly sexual activities of the daughters and the *qĕdēšôt* in Hosea 4:13–14 to cultic prostitution[30] or to a fertility ritual described by Herodotus.[31] Yet, these translations and analyses convey a gender bias: They interpret the "whoring" of the entire people in v. 12 metaphorically as false worship, but the "whoring" of the female subjects in v. 14 literally assuming that women and female cult servants would be engaged in sexual intercourse at the sanctuary.

In contrast, I argue that all renderings of the verb *znh* (to play the whore) in Hosea 4:11–14 do not name actual sexual rites but metaphorically condemn any ritual at the open-air *bāmôt* as idolatrous with regard to Israel's national deity.[32] The notion of female worshippers and cult personnel in this passage may indicate that women are particularly active in this cult that may involve rituals for Asherah or some other female deity. A peculiar feature of the indictment is, however, that it addresses the men who seem to take part in the sacrifice and who in a patriarchal society are supposed to control the women of the family. Thus, the polemic of Hosea 4:11–14 does not reproach just one group (women) or social class (cult personnel) in the Northern kingdom but actually condemns the whole people (v. 12a, 14b). The prophetic voice legitimizes its dominance by citing Israel's national deity (v. 12a, 14a) and therefore does not allow the alternative view, which can only be reconstructed by inverting the polemics.

Conceptualized Space or Conceived Space

Since the cult described in Hosea 4 is local and the sanctuaries are numerous, the prophet refers to the concept of the *bāmôt* as sacred spaces, that is, contact zones to the divine, that link the rural population to the land and its fertility. My assumption that the sacred spaces condemned by Hosea are such local sanctuaries at which fertility deities are venerated is strengthened by Hosea 2:4–15, which offers a monologue of Israel's national deity in the role of a jealous "husband" who accuses his "wife" (the Israelite people) of following her "lovers" (other deities) in the hope that they will provide nourishment and clothing. The personified people is cited in v. 7, "I will follow my lovers, who give me my bread and my water, my wool and my flax, my oil and my drink" to which God, her husband responds, "she did not know that I myself gave her the grain, the must, and the freshly pressed oil; therefore I will take back my grain in its time and my must in its season; I will snatch my wool and my flax, which were to cover her nakedness" (v. 10–11). Bread and oil, wool and flax, are the products of an agrarian culture and the basic means for survival. In the hilly countryside of the Northern kingdom, the growth of grain and olives is highly dependent on winter rain. Hosea 2:4–15 thus reveals the conflict between the people who venerate deities of fertility and the prophet who portrays the Israelite deity as the provider of food and clothing in the role of the true "husband" of Israel whose basic duty is to care for his wife.

Spaces of Representation or Lived Space

By reflecting on the third perspective on space—space as experienced by users who attribute symbolic value to a place—the dissent between the addressees and the dominant voice in Hosea 4:11–14 becomes even more evident. The note that the people seek "good" shade under the trees (4:13a) reveals the religious significance of the local sanctuaries to the people, since a lush green tree not only offers shade, but symbolizes nourishment and protection as well, which are often linked to female deities. Silvia Schroer has convincingly demonstrated that in Syria-Palestine a green tree symbolizes a goddess, which on Middle Bronze Age seals is depicted nude with twigs or trees, often *en face* or with enormous ears.[33] This image is a special case of the iconography of the Syrian *dea nutrix*. While nudity symbolizes vitality and eroticism, the tree or twigs connote fertility, and her face and ears attention. In Palestine the tree eventually substituted the goddess following the general demise of anthropomorphic representations of deities since the Late Bronze Age. On seal impressions from different locations of the Northern kingdom (Samaria, Megiddo, and Bethel), which date to the Iron Age (1200–700 B.C.E.), the tree is venerated by a human figure.[34]

Marie-Theres Wacker rightly points out that the Hebrew term *ʾēlâ* (terebinth, Hosea 4:13) is homonymous to the feminine form of *ʾēl* (God).[35] Following Schroer she interprets the mighty trees in Hosea 4:12–13 as references to a goddess.[36] In eighth-century Israel this goddess was most probably called Asherah, her domain being the fertility of fields, herds, and humans. According to 1 Kgs 16:32–33 the Israelite King Ahab built an altar for Baal and made "the Asherah," probably a cult statue of the goddess.[37] The connection between this goddess and a wooden pole or tree trunk is attested in the exilic polemic against the "Asherim" in the Deuteronomistic history.[38] The biblical references to Asherah are confirmed by extra-biblical sources. In the southern Negev, formulas of blessing from the caravan station of *Kuntillet ʿAjrud* and from a tomb inscription of *Khirbet el-Qôm* mention a god named "Yahweh of Samaria" and "Yahweh of Teman" together with "his Asherah."[39] Although these formulas of blessing document personal faith, they witness the veneration of a local Yahweh figure with his female consort Asherah.

Within this religious context, it is highly probable that a divine couple supposed to protect the towns and provide fertility for the farmland was venerated at the *bāmôt*. The standing stone (*maṣṣēbâ*) and the tree are their symbols. It is not clear, however, whether the rituals at the local sanctuaries were dedicated to Israel's God and his consort or to some other deity besides Asherah. Both the Syrian weather-god Baal and a god named Yahweh are attested as consorts of Asherah, and they share a common symbol, the calf.[40] The texts in Hosea do not provide sufficient distinctive information and even conceal the name of the goddess. While the prophetic polemic in Hosea 4:11–14 strictly separates Israel's main deity from the *bāmôt*, the portrait as provider of agrarian produce in Hosea 2:4–14 points to the opposite direction. In Hosea 14:8 Israel's God even replaces the goddess, when he is likened to a luxuriant tree that delivers fruit to Ephraim. Obviously, this verse transfers the symbolism attached to the unnamed goddess to Israel's main deity.

This reconstruction of the rituals at the *bāmôt* suggests that the rural population of the Northern kingdom venerated a divine couple, probably named Yahweh and Asherah, at these sites. In contrast, Hosea argues that Israel's true deity, his God, is not like other deities and that the goddess has to be dismissed. The basic argument of Hosea and his followers is that God's competence to provide food and clothing for Israel has nothing to do with any gendered personality or competence as a weather-god, but with his role as husband of Israel. The book's leading concept of a marital relationship between God and female Israel excludes all relations to other deities.

In sum, the conflict about the *bāmôt* in Hosea is one about *lived* space, the religious symbolism of the local sanctuaries and their social significance for the Israelites. The prophet polemically dismisses these sites because their religious symbolism referred to a divine couple responsible for the fertility of the land. Consequently, an oracle of doom even announces the "undoing" of sacred space:

> The *bāmôt* of Aven, the sin of Israel, shall be destroyed.
> Thorn and thistle shall grow upon their altars.
> They shall say to the mountains, "cover us," and to the hills, "fall on us." (Hosea 10:8)

If thorns and thistle overgrow the altar and the hills cover people and settlements, the *bāmôt* forfeit their significance with regard to both spatial practice and representation of space. While this verse does not tell who destroys the local sanctuaries, the context presents the plundering of Samaria by the Assyrians (10:6) as God's punishment (10:10–12). According to the authors of Hosea 10 and certainly for people living after the collapse of the Northern kingdom, it was Israel's God who destroyed the local sanctuaries. At this point, the question arises where the legitimate sacred space would be.

The House of God at Bethel

One would think that Hosea may promote the royal or state sanctuary at Bethel as a legitimate sacred space. The use of the pejorative name Beth-Aven "house of disaster" in Hosea 4:15; 5:8 and 10:5, however, devalues even this sanctuary. Hosea 4:15 indicts Israel of "playing the whore" and therefore forbids pilgrimages to Beth-Aven and oaths in the name of Israel's main deity. The verse firmly connects Israel's veneration of other deities to a loss of communication with the "true" God. In Hosea 5:8, Beth-Aven is listed among other towns in the region of Benjamin in which people will be called to arms. Lastly, Hosea 10:5 mentions the sacred object of Bethel's sanctuary in a highly negative outlook.

> *5* The inhabitants of Samaria are worried about the calfling of Beth-Aven.
> Indeed, its people already mourn over it and its priestlings cry loudly,
> about its glory, because it has been taken from it.
> *6* It also shall be carried to Assyria as tribute to the great king.
> Ephraim will receive shame and Israel will be ashamed of his counsel.

This prophetic reflection about the Bethel sanctuary summons the end of the Northern kingdom in the near future. The calf of Bethel, that is, the image that once represented Israel's national deity, will be carried to Assyria like many other precious things. The calf, in Hebrew *ʿegel*, is pejoratively named *ʿeglot* (calfling or calf-like-thing). The term used for the priests of the sanctuary appears elsewhere only in polemical contexts (2 Kgs 23:5; Zeph 1:4) and is thus rendered with "priestlings" in the above translation. Therefore, even the royal sanctuary at Bethel and its cult image of the calf are said to be doomed. From the perspective of the dominant prophetic voice, there is no legitimate sacred space for Israel.

Conclusion

Analyzing the polemic against the *bāmôt*, the local sanctuaries, and the negative outlook on the royal sanctuary at Bethel and its cult object, it is obvious that Hosea and his followers try sharply to distinguish Israel's God from other deities and even from the image of the calf that was formerly connected to him. The dominant voice in Hosea portrays Israel as an ignorant people who do not realize their God's competence in providing grain, wine, and oil, the fruits of the land. With the help of Lefebvre's trialectic theory of space, Hosea 4:11–14 and other texts can be interpreted as witnessing a spatial practice of the Israelites in the Northern kingdom. Through cultic installations, sacrifice, and rituals of divination they establish local sanctuaries as sites of contact to a divine couple thought to safeguard their towns and the fertility of their farmland. The prophetic voice, however, dismisses these sacred spaces and their popular cult as illegitimate. This voice opposes the social significance of the local sanctuaries by arguing that Israel's use of these sites and their assumed symbolism compromises the exclusive relationship of Israel to its national deity and even jeopardizes the royal sanctuary at Bethel. In light of the metaphor "God is the husband of Israel," which means that he is *not* the husband of the goddess Asherah, the rituals at the *bāmôt* are evaluated as "whoring away from beneath their god" (Hosea 4:12). This indictment of the sacred spaces is, however, clearly in opposition to what the Israelites thought and pursued in their spatial practice.

Notes

1. Henri Lefebvre, *La production de l'espace* (Paris: Éditions Anthropos, 1974); *The Production of Space,* D. Nicholson-Smith, trans. (Malden, MA: Blackwell, 1991). The following references and citations are taken from the English edition. For an evaluation of Lefebvre's theory on space and its influence on other fields see Edward W. Soja, *Postmodern Geographies: The Reassertion of Space in Critical Social Theory* (London: Verso, 1989), esp. 43–75; Jon L. Berquist and James W. Flanagan, "Ancient Perceptions of Space/Perceptions of Ancient Space," *Semeia* 87 (1999): 15–43.
2. Cf. Lefebvre, *Production*, 33.
3. Cf. Lefebvre, *Production*, 33, 38.
4. Cf. Lefebvre, *Production*, 40.
5. Lefebvre, *Production*, 38. Lefebvre uses the term *representation* as a broader notion that subsumes the area where *knowledge* and *ideology* are barely distinguishable, cf. *Production*, 45.
6. Cf. Lefebvre, *Production*, 40.
7. I prefer this term to the translation "representational spaces" used by D. Nicholson-Smith; the French reads "*les espaces de représentation.*" I thank Mark George for discussing Lefebvre's theory, which helped me to modify my perception of both the collective and the individual level of the three dimensions. In his book *Israel's Tabernacle as Social Space* (Atlanta, GA: Society of Biblical Literature, 2009), 23–27, George uses the terms "spatial practice," "conceptual space," and "symbolic space" for Lefebvre's triad.
8. Lefebvre, *Production*, 39.
9. Cf. Jonathan Z. Smith, "The Topography of the Sacred," in *Relating Religion: Essays in the Study of Religion* (Chicago: University of Chicago Press, 2004), 101–16, 110.
10. Jonathan Z. Smith, *To Take Place: Toward Theory in Ritual* (Chicago: University of Chicago Press, 1987), 96–117; Smith, "Topography," 108–11.
11. For the socialhistorical context of the female personification of city and land see my study Christl M. Maier, *Daughter Zion, Mother*

Zion: Gender, Space, and the Sacred in Ancient Israel (Minneapolis: Fortress Press, 2008), 61–74.

12. Cf. Maier, *Daughter Zion*, 21–28.
13. Cf. Lefebvre, *Production*, 172–74.
14. The number of articles and books on this issue is overwhelming. For an introduction see Gerlinde Baumann, *Love and Violence: Marriage as Metaphor for the Relationship between YHWH and Israel in the Prophetic Books,* Linda M. Maloney, trans. (Collegeville, MN: Liturgical Press, 2003); Alice A. Keefe, *Woman's Body and the Social Body in Hosea,* JSOTSup 338 (Sheffield: Sheffield Academic Press, 2001); Yvonne Sherwood, *The Prostitute and the Prophet: Reading Hosea in the Late Twentieth Century* (London: T&T Clark, 2004).
15. Verbal forms of *znh* occur in Hosea 1:2 (twice); 2:7; 3:3; 4:10, 12, 13, 14 (twice), 15, 18 (twice); 5:3; 9:1; the abstract plural *zĕnûnîm* in Hosea 1:2 (twice); 2:4, 6; 4:12; 5:4. Out of twenty occurrences of the root *znh* in the book, nine are clustered in Hos 4. Forms of *n'f* are used five times in Hosea (3:1; 7:4), among them three occurrences in Hosea 4:2, 13, 14.
16. Cf. Phyllis A. Bird, "'To Play the Harlot': An Inquiry into an Old Testament Metaphor," in *Missing Persons and Mistaken Identities: Women and Gender in Ancient Israel* (Minneapolis: Fortress Press, 1997), 219–36.
17. 1 Kings 12:25–29 offers the foundation legend of both royal sanctuaries at Bethel and Dan by Jeroboam I. Bethel is mentioned 44 times in the Hebrew Bible and is "the king's sanctuary and a temple of the kingdom" (Amos 7:13). For a full assessment of the Bethel tradition cf. Klaus Koenen, *Bethel: Geschichte, Kult und Theologie,* OBO 192 (Freiburg: Universitätsverlag, 2003).
18. Hans Walter Wolff, *Dodekapropheton 1: Hosea,* BKAT 14/1, 3rd ed. (Neukirchen-Vluyn: Neukirchener, 1976), 93 sees in Chapters 4–12 sketches of Hosea's oracular performance noted down by his disciples close to the events. Similarly Francis Andersen and David N. Freedman, *Hosea: A New Translation with Introduction and Commentary,* AB 24 (Garden City, NY: Doubleday, 1980), 52–53; James L. Mays, *Hosea: A Commentary,* OTL (Philadelphia: Westminster Press, 1969), 5–7.
19. Jörg Jeremias, *Der Prophet Hosea übersetzt und erklärt,* ATD 24/1 (Göttingen: Vandenhoeck & Ruprecht, 1983), 18 views the book as a product of a later (Judean) redaction, yet he also assumes that Chapters 4–11 convey the prophet's words written down soon after the demise of the Northern kingdom.
20. So with Mays, *Hosea*, 72–73; Jeremias, *Hosea*, 68–69.
21. The concentric structure is also argued by Jeremias, *Hosea*, 68. The structure of verses (a, b) in my translation follows the Masoretic text (= MT), that is, the original Hebrew consonantal text as vocalized and accented by the Masoretes (sixth to ninth centuries CE) and edited

by the Deutsche Bibelgesellschaft in Biblia Hebraica Stuttgartensia (= BHS), Stuttgart 1967.

22. My literal translation is intentional as it marks an important locative aspect, which will be discussed below.
23. Cf. Deut 7:5, 1 Kgs 14:23; 2 Kgs 23:15; Mays, *Hosea*, 74; John Day, "Asherah," *ABD* 1:486. A *maṣṣēbâ* is also mentioned in Hosea 3:4 in a list together with sacrifice and ephod, and next to altars in Hosea 10:1–2.
24. Matthias Gleis, *Die Bamah*, BZAW 251 (Berlin: De Gruyter, 1997), 246; Similarly Klaus-Dietrich Schunk, "*bāmâ*," *TWAT* 1:663–66.
25. Cf. Gleis, *Bamah*, 247–48. The Deuteronomistic editors of Israel's narrated history (Deuteronomy–2 Kings) condemn all worship outside of one divinely chosen sanctuary (cf. Deut 12, supposedly referring to the temple of Jerusalem) and therefore polemically dismiss the *bāmôt* in 1 Kgs 12:31; 14:23; 2 Kgs 17:9, 11, 32; 23:8.
26. For a fuller treatment of the issue see my article Christl M. Maier, "Myth and Truth in Socio-historical Reconstruction of Ancient Societies: Hosea 4:11–14 as a Test Case," in *Thus Says the LORD: Essays on the Former and Latter Prophets in Honor of Robert R. Wilson*, J. J. Ahn and S. L. Cook, eds., LHBOTS 502 (New York: T&T Clark, 2009), 256–272.
27. Cf. Sir James Frazer, *The Golden Bough: A Study in Comparative Religion*. 2 vols. (London: Macmillan, 1890). Especially the abridged English version, published in 1922 and translated into German in 1928, proved to be most influential since the book was an integral part of private libraries. Cf. Julia Assante, "From Whores to Hierodules: The Historiographic Invention of Mesopotamian Female Sex Professionals," in *Ancient Art and its Historiography*, A.A. Donohue and M.D. Fullerton, eds. (Cambridge: Cambridge University Press, 2003), 13–47, esp. 23–24.
28. Cf. Assante, "Whores," 19–26; the reception of Frazer's theses among theologians is commented by Christine Stark, *«Kultprostitution» im Alten Testament? Die Qedeschen der Hebräischen Bibel und das Motiv der Hurerei*, OBO 221 (Fribourg: Academic Press, 2006), 29–43.
29. The term is rendered "cult prostitute" (RSV), "temple prostitute" (NRSV, cf. "Tempeldirne" in the German Luther translation [revision of 1984]), "sacred prostitute" (New Jerusalem Bible, cf. "Weihedirne" in Einheitsübersetzung).
30. Cf. Mays, *Hosea*, 75; Jeremias, *Hosea*, 71; Andersen and Freedman, *Hosea*, 368.
31. Cf. reference to Herodotus, *Histories* I 199 in William R. Harper, *Amos and Hosea*, ICC (Edinburgh: T&T Clark, 1994 [1st ed., 1904]), 261; Wolff, *Hosea*, 108–9. New studies on Herodotus reveal that he often exaggerates and talks derogatorily about the sexuality of other peoples, cf. Stark, *Kultprostitution*, 12.

32. Cf. Maier, "Myth," 266; for a similar evaluation cf. Karin Adams, "Metaphor and Dissonance: A Reinterpretation of Hosea 4:13–14," *JBL* 127 (2008): 291–305.
33. Silvia Schroer, "Die Zweiggöttin in Palästina/Israel: Von der Mittelbronze II B-Zeit bis zu Jesus Sirach," in *Jerusalem: Texte–Bilder–Steine. Festschrift Hildi Keel-Leu und Othmar Keel,* M. Küchler and C. Uehlinger, eds., NTOA 6 (Freiburg: Universitätsverlag, 1987), 201–25; cf. also Schroer, "Die Göttin auf den Stempelsiegeln aus Palästina/Israel," in O. Keel, H. Keel-Leu and Schroer, *Studien zu den Stempelsiegeln aus Palästina/Israel II,* OBO 88 (Freiburg: Universitätsverlag, 1989), 89–207.
34. Schroer, "Zweiggöttin," 212–15; cf. also Othmar Keel and Christoph Uehlinger, *Gods, Goddesses, and Images of God in Ancient Israel,* Thomas H. Trapp, trans. (Minneapolis: Fortress Press, 1998), 94–95.
35. Marie-Theres Wacker, *Figurationen des Weiblichen im Hoseabuch,* HBS 8 (Freiburg: Herder, 1996), 275–76.
36. Cf. Wacker, *Figurationen*, 282–90.
37. Cf. also 2 Kgs 13:6. 1 Kings 15:13 and 2 Kgs 21:3 mention that her image, that is, her statue, was even placed in the temple of Jerusalem.
38. Cf. Deut 7:5; 12:3; 2 Kgs 18:4; Josh 6:25–30 mentions Baal's altar and the tree symbolizing Asherah.
39. Among numerous studies on the finds of *Kuntillet 'Ajrud* and *Khirbet el-Qôm* cf. André Lemaire, "Les inscriptions de Khirbet el-Qôm et l'Ashérah de YHWH," *RB* 84 (1977): 595–608; William G. Dever, "Asherah, Consort of Yahweh? New Evidence from Kuntillet 'Ajrud," *BASOR* 255 (1984): 21–37; Keel and Uehlinger, *Gods,* 128–47. Kuntillet 'Ajrud was probably a trading post of the Northern kingdom.
40. Cf. Mark S. Smith, *The Origins of Biblical Monotheism: Israel's Polytheistic Background and the Ugaritic Texts* (Oxford: Oxford University Press, 2001), 32, 56–57, 67–80.

Chapter 7

Mapping Our Last Places: Apocalyptic Space and Imagery at Chartres Cathedral—A Social and Visual Analysis of Imagined Space

Katherine Rousseau

The very act of representing our imagined spaces may contribute to social patterns and practices in our lived spaces. Notably, this is the case for the charged spaces where we imagine and represent apocalyptic events. Thompson notes an increased attention to the spatial aspects of apocalyptic imagination in recent scholarship: "In the study of apocalypses, time has passed and space is taking its place; at least, space is taking a place alongside time."[1] This emerging focus is shown through explorations of climate, geography, astronomy, meteorology, and spatially oriented elements in apocalyptic discourse.[2] Texts such as 1 Enoch, the Ascension of Isaiah, and the Apocalypse of Paul focus on cosmological secrets or otherworldly geography, as the visionary tours the levels of heaven, the zones of hell, and in some cases, the architecture of the heavenly city or temple. More recently, scholars have examined the broader social implications of apocalyptic ideas by means of critical approaches to space, with studies directly engaging with the rhetorical and discursive aspects of apocalyptic space.[3]

As these studies demonstrate, apocalyptic ideas push us to be conscious of space, if less directly than they push us to be conscious of time. We are invited to consider the architecture of the cosmos and eschatological scenarios, all unfolding in some anticipated space at some anticipated time. Moreover, when we attempt to visually represent such imagined or anticipated space, our images serve as a delivery system for certain spatial ideas. As we depict otherworldly territories, we also situate the places of our own world in a temporal-spatial plan.

As Himmelfarb states, expressed in apocalyptic texts is a belief that the boundary between the human and the divine is somehow permeable.[4] Likewise, critical approaches to apocalyptic space may explore the permeability of boundaries between human and divine territories, as well as the relationship between our imagined territories and the lived spaces of our own world.

As a particular case study in apocalyptic visualization, I will assess the functions of the apocalyptic space presented to the medieval spectator and pilgrim at the south transept of the Chartres cathedral in France. My particular focus is the south transept's central portal tympanum depicting the Last Judgment. The first part of my chapter is concerned with ideas about space in relation to the images presented in the Last Judgment tympanum—the textual resources, cosmological blueprints, and theological priorities put forward by the imagery. The second part is concerned with spatial practice at the south transept, focusing on the thirteenth century: the built environment, the daily routines and pilgrimage routes, and the economic and social zones in proximity to the south porch.[5] An examination of the interplay between abstract ideas about space and spatial practices provides a critical assessment of apocalyptic space at work in a particular context. Such a study of the Chartres demonstrates how representations of imagined space may acquire specific rhetorical, social, and political significance. That is, the depiction of judgment at the portal was a declaration not only about the nature and structure of the cosmos, but also about the community space of Chartres itself.

Spatial Concepts and Last Judgment: Text, Cosmology, Theology

To the contemporary eye, the Cathédrale Notre-Dame de Chartres and its surroundings are social space with variegated purposes: people capture pictures with their video phones and make purchases in the souvenir shop outside the south transept. Aside a few obvious markers—clothing such as the nun's habit or banners held by small groups entering through the south doors—it is difficult to tell who has come to Chartres for what purpose: one comes as a tourist to see the UNESCO World Heritage site as an architectural and cultural landmark, while another comes as a present-day pilgrim following the route to Santiago de Compostela. As the stage for this varied activity, the building itself still rises above the town, a dominating feature of the landscape: the cathedral was constructed on the highest ground of the small city of Chartres, on the left bank of the River Eure, and is surrounded by fertile plains. The structure was substantially rebuilt in a three-decade period after a fire in 1194 and consecrated in 1260. Now, we consider the cathedral emblematic of French Gothic style: its light, verticality, and basilica design, as well as its vaulted ceilings, buttresses, and large stained-glass windows offer a checklist of quintessential Gothic forms and ideas.

As part of the sculptural program of the cathedral, the south façade (ca.1220) presented a range of possibilities to the medieval spectator: the sculptural façade taught lessons, chronicled history and commemorated events, and elicited emotional or memory responses. At the Chartres, the porch sculpture was commissioned by the cathedral chapter itself, with iconographic content planned by the clerics.[6] The sculpture at the south transept featured a familiar scene for medieval portal programs, the Last Judgment. This scene was pivotal: it was the intricately carved central portal tympanum, flanked by the side portal tympana depicting the martyrs and the confessors (figures 7.1 and 7.2). In the central tympanum, Christ is depicted as the enthroned cosmic judge, displaying the marks of crucifixion, between Mary and John the Evangelist. Mary and John hold their hands together, faces turned toward Christ; they appear as intercessors, pleading for the dead. Below Christ, on the lintel and the lower

Figure 7.1 South transept, Chartres Cathedral.

Source: Photo: Stephen O. Gombosi.

Figure 7.2 Last Judgment, center portal tympanum of south façade, Chartres Cathedral.

Source: Photo: Stephen O. Gombosi.

archivolts, the dead awaken, emerging from their tombs in clusters. The archangel Michael stands on the lintel at the midpoint of the dead, holding scales to weigh souls and dividing the line into two sections, the blessed and the condemned. The angels assemble in the arch above, while the hellmouth and its demons are on Christ's lower left, or the viewer's right.

There are a number of abstract ideas about space informing this scene, competing and converging. By the thirteenth century, Last Judgment imagery derived from accretions of textual references, theological concepts, and more loosely assembled codes; these became associated in Christian iconography with the last events of this world and the transition to a new and purified place. One textual source for these events is Revelation 20: this chapter of the Christian Bible narrates the final fate of the dead, who rise from their sleep at the end of the world to be judged on their actions and assigned their ultimate place in either heaven or hell. Christ is described seated on a white throne of judgment in some space unrecognizable to us, where "earth and the heaven fled from his presence, and no place was found for them" (Rev. 20:11 NRSV). The natural environment has contained souls until this time of judgment, when the sea and the underworld territories surrender the dead. The tympanum scene engages with this passage, even as it features additional eschatological themes, such as the Virgin Mary as intercessor for the dead.

The theological and artistic deployment of this scene was not unique to Chartres. Similar iconography was used in the twelfth and thirteenth centuries' Last Judgment tympana at Saint-Denis, Autun, Conques, and Bourges: the central zone of Christ enthroned; the horizontal registers creating bands of activity; and the dead emerging from their own tombs.[7] As Emile Mâle points out, Last Judgment imagery employed at thirteenth-century churches and cathedrals flowed from a complex mix of text, theology, and popular fancy. During this historical period, the Last Judgment was understood as a great drama, and the sculptor's challenge was to present the chronologically unfolding story in a single artistic space.[8] As part of this challenge, the craftsman had to convey the stuff of visions, allude to the architecture of heavenly sites, or shorthand the territories of the otherworld, all necessitating some reference to spatial conditions.[9]

Adding to these textual and iconographic ideas, a cosmic blueprint underpins the tympanum scene. The medieval scenario of final judgment employed a spatial scheme that mapped the contrasting regions of earth, heaven, and hell. In his study of apocalyptic space, Barasch notes that Last Judgment scenes, as articulations in the medieval imagination of the final human place in cosmic history, incorporated specific spatial structures. We can locate three principal patterns: first, the generation of separate regions through "space-creating devices," such as the halo or mandorla; second, the hierarchical structuring of space into upper and lower levels, where the dark and malignant underworld spaces are placed below the crust of the earth and heaven is located in the celestial sphere above earth; and third, the horizontal dividing of space into left and right.[10] These codes of orientation would be familiar to the medieval observer—who would know, for example, that condemned souls are located on Christ's left, both from visual precedent and from narratives like Matthew's parable of the sheep and the goats.[11] These structures allowed the spectator to recognize the sinners and the blessed according to the space reserved for them. These tactics organized space into an intricate and orderly topography: areas and directions had their own character and as such, meaning could be derived from the particular characteristic of the area or direction.[12]

Within these spatial structures, representations of the body served as a teaching tool, conveying important lessons for the medieval spectator: the female body of Mary with a pleading posture, for example, or Christ as cosmic judge exposing the skin of his chest and revealing the physical wounds that marked crucifixion. The dead are embodied, clothed, and emotional, hands clasped hopefully. In the line of souls to be judged, expressions instruct and forewarn: at the outer edges of the lintel, two blessed souls clasp each other's hands, while the last outer figure in the line of the condemned looks out towards the viewer's space, stricken, as the orderly line gives way to a chaotic heap of demons at the hellmouth.

Notably, the positioning of bodies in this spatial scheme has a constructive effect: a body becomes defined by its symbolic placement, and that definition gives information about these otherworldly territories.[13] The demons in the tympanum scene, for example, have human torsos and limbs, but also horns, animal ears, and muzzles—a characteristic of Gothic imagery, where stone creatures with distorted faces, teeth, and animal features participated in apocalyptic events.[14] This signals to the viewer that the creatures of each region possess bodily characteristics corresponding to the space. The space at the hellmouth is designated for those creatures with the physiognomic signs of vice, in contrast to the angels of the heavenly region above, whose more serene expressions are the physiognomic signs of virtue.[15] That is, the space can be understood by examining the nature and position of the bodies represented within it: the body both conforms to the space and expresses the nature of the space.

Such textual resources and cosmological maps provided a set of spatial inflections for the south transept central tympanum scene. These ideas were universalizing; they were not specific to Chartres and they incorporated broad theological debates about eschatology, body, and salvation. At the same time, the spectator encountered the visual representations of apocalyptic space in the particular social space at the south porch. The portal sculpture was not simply a self-contained pictorial network referencing external themes and texts; rather, the artistic program participated in the spatial order at Chartres. In order to understand how these representations of apocalyptic space functioned, we must examine this social space: the built environment of the cathedral site and the spatial practices at and around the south transept.

Spatial Practices and Symbolic Resources at the South Transept

The European cathedrals undertaken in the thirteenth century were structures of immense power: art historian Michael Camille has called them "vast mass-machines" that were "not unlike the shimmering

Postmodern towers of today's corporate headquarters."[16] The cathedral site was a linchpin in the intricate apparatus connecting city and cosmos in the European medieval worldview: the urban landscape was a microcosm, where all territories were oriented in a revealed plan for history, underscoring the geographic centrality of the community's main church.[17] Everything had its assigned place in this plan, from the highest abstraction of heaven to the local places of worship and commerce. The question "where am I?" had a local and social answer that was, in turn, deeply embedded in a cosmological answer.

Consequently, a cathedral told a comprehensive story about its own position. The building not only narrated its place in sacred history, it also subsumed the daily spaces of everyday life. Stone and glass chronicled the arc of time since creation and the Hebrew prophets, cataloguing phenomena such as the virtues and vices, the zodiac signs, and the labors of the month. Certainly, at Chartres, the cathedral dominated the town in a dramatic way, embedding itself in local practices. Its physical presence high on the landscape and the scope of the building project assured its political and economic centrality, as well as its cultural and aesthetic importance.[18] The expense of the site included costly media of sculpture and stained glass, intended to impress the onlooker and pilgrim, demonstrating the place's venerability and political power.[19] The cathedral at Chartres, then, marshaled the particularity of its own grandeur to emphasize its importance in the divine cosmic plan.

The Last Judgment portal and its apocalyptic imagery provided a key landmark in this cosmic plan, orienting various social activities and declaring a relationship between the cathedral and the community. In viewing the tympanum scene at the portal, the spectator was positioned at the point of entry to the cathedral's interior salvific space—as Olson observes, the church portal is both catalyst and referent, serving as the boundary that defines the state of salvation.[20] In the process, the sculpture's imagery aimed to evoke memory—that is, according the period's visual idiom, an image was important because of its theological significance and its power to remind the viewer of an event, textual passage, or idea. The wounds of crucifixion displayed by Christ in the scene, as one instance, reminded the devotee of both the end of life's pilgrimage and the end of time, conveying the urgency of repentance.[21] At the doorway, then, the tympanum's ability to prompt recognition and emotion was more important than optical illusions of volume or realistic spatial relationships between objects to unify the composition.[22] In this sense, the apocalyptic space presented in the tympanum was created according to historically situated spatial codes, both artistic and social, meant to engage the spectator in the activity of seeing.

This activity of seeing presented multiple possibilities for engagement: events in the tympanum scene, for example, were made tangible to the audience through the deployment of realistic figures in its apocalyptic space. These were figures with identifiable and symbolic social roles, represented through costume. This includes a court official with a circlet of flowers in his hair in the line of the blessed on the lintel.[23] As well, on the lower archivolt to Christ's left, the demons escort an aristocratic woman, a nun, and a wealthy man clasping his moneybag to his chest. Hollengreen characterizes this scene as an incipient medieval Dance of Death, where the careful rendering of different costumes displays a range of social identities to the spectator: regardless of sex, class, or profession, everyone is subject to the eschatological drama.[24] Thus, not only does the tympanum incorporate a sacred geography and sacred history, it also offers a kind of sociology of medieval society.[25]

The activity of seeing and its associated social symbolism become especially important when we locate Chartres as a pilgrimage site and Marian shrine. The nature of medieval pilgrimage to Chartres, centering on the relic of the Virgin's tunic, continues to be debated by scholars. According to Bugslag, pilgrimage and Marian devotion at Chartres varied according to social class and time period; activity ranged from twelfth-century mass pilgrimages, such as those petitioning Mary for help with frequent outbreaks of ergot poisoning, to military pilgrimage by knights eager to touch their chemises to the Sainte-Châsse reliquary, the container for the tunic, to secure protection and victory.[26] Bugslag draws upon the work of Alphonse Dupront for his analysis: Marian sites had their own individual character, but functioned within a prescribed set of elements that included social, spatial, and environmental aspects.[27] Thus, the Virgin Mary and her compositional prominence in the Last Judgment tympanum reflected local Marian worship: the relatively large size of the figure of Mary and her elevation to the same register as Christ emphasize her overall importance at Chartres—what Spitzer calls a "locally specific" version of Mary.[28] At the same time, this prominent display of Mary as intercessor in the tympanum scene also participated in a more general shift in apocalyptic iconography. By the thirteenth century, tactics for communicating eschatological lessons to the viewing audience were changing, and Last Judgment imagery had transitioned from an emphasis on divine wrath toward a more hopeful expression of intercession and salvation. As Williamson points out, the church needed to connect with audiences in different ways.[29] As such, the tympanum's imagery participated in broader iconographic developments that were filtered through the site and its devotional cult at Chartres.

The tympanum subject matter displayed the site's importance for the cosmic plan and for the community. This importance extended beyond the devotional practices supported by the tympanum's iconography and incorporated the broader economic and social life operating outside the south transept. The Last Judgment portal was part of an architectural address to the town itself. As Hollengreen argues, different portals at Chartres served different functions for various audiences in the city: while the grand west façade was the main liturgical entrance, the south façade was a major point of access for the public, and the principal urban constellation, with its commercial and artisanal activities, lay to the east and south.[30] Scholars have also noted the proximity to the castle of the counts of Chartres and their sphere of influence, directly to the south of the cathedral.[31] Thus, on the doorstep of the south transept, with its Last Judgment portal, was a dividing line between competing feudal lords—the bishop in his cathedral and the count in his castle.[32]

Notably, the south transept façade is, as an architectural ensemble, visually commanding, towering over the landscape in a way that the north façade does not. Even as it signaled the site's role in the cosmic plan and sacred history, the Last Judgment tympanum also affirmed church authority within the city of Chartres. Says Hollengreen:

> The sculptural programme of the south façade, comprising a Last Judgment portal flanked by portals dedicated to Martyrs and Confessors, appears to have been designed with address to this urban setting in mind. Of all the High Gothic portals at Chartres, the Last Judgment portal is the most broadly relevant to potential viewers—it forcefully and explicitly comprehends *every* viewer (male/female, cleric/layperson, knight/bourgeoisie, king, bishop, monk, and maiden) in a way none of the other imagery does—and it achieves the broadest possible public address by virtue of its position facing the largest concentration of urban population.[33]

Williams adds that the south porch served very practical purposes in the town: it faced the largest exterior open space where lively cathedral fairs were held.[34] These fairs, associated with the feasts of the Virgin, provided an important economic zone in the space outside controlled by the cathedral chapter. The sale of religious souvenirs and devotional objects to visiting pilgrims was key, and tradesmen and guilds were deeply involved in the activity. Fuel, vegetables, and meat were the main commodities sold outside the south portal.[35] Money-changing tables were also installed—and the close proximity of the carved statue displaying the personification of Greed on the south porch, notes Williams, may not have been a chance occurrence.[36] In particular, expenditures by pilgrims generated significant income for the townspeople of Chartres.[37] Items for sale included badges and chemisettes for expectant mothers, as well as the small lead pendants of the early thirteenth century bearing the image of Mary's veil.[38] In these ways, the order of the cosmos encompassed an economic order, unfolding in the space outside the south portal.

The Last Judgment portal's apocalyptic content raised the stakes through its broad public and social address: it located the inevitability of final judgment and then gave way to the salvation inside—a salvation only the church could offer. This was a strong declaration, physically juxtaposed with the commercial and aristocratic power directly outside. Certainly, this declaration carried with it certain tensions: the south transept façade incorporated both bishops in the tabernacle nave buttresses and kings in the tabernacles between the south porch pediments—flanking the Last Judgment tympanum, kingship and priesthood were presented in utopian harmony. This sculptural display of both *regnum* and *sacerdotium* stressed not only the unity and balance between the two powers, but "an idealized image of the whole social order," comprising groups with complementary but different roles and statuses.[39] At the same time, however, we can point to evidence that such harmony at the local level was more idealized than actual: the violent dispute between Chartres and the count's vassals in 1210, for example, or the competing systems of taxation on vassals established by count and church.[40] The sculpture's harmonious imagery, then, modeled social order in the community, even as it used the purposeful machinery of the cosmic order to explain the church's supreme role in the city. Images of kingship appeared in the south portal sculpture program, but Christ as final judge held the pivotal apocalyptic role at the tympanum above the central doorway, high symbolically and compositionally, offering a commentary on local power. This power, embodied on earth by the bishop and his canons, was the highest and last authority.

Conclusion: This-Worldly Implications of Imagined Spaces

The Last Judgment tympanum at Chartres deployed the codes of imagined apocalyptic space. Simultaneously, the site where the tympanum was displayed was itself a social institution, oriented toward urban political power and economic activity—a palpable terrestrial manifestation of heavenly order and splendor. The case of Chartres demonstrates how apocalyptic space as a category can be examined through its representations in particular places: this not only allows for an assessment of the variations and developments in iconography, it also invites us to consider how these representations functioned in a community. As presented in the tympanum, apocalyptic space, with its carefully ordered cosmology, derived additional meaning from local spatial practices at the site of display. Consequently, apocalyptic space at the Chartres south portal was not exclusively an *abstract* concept, portraying a general set of eschatological ideas; rather, representations of imagined apocalyptic spaces participated in specific social relations and agendas. These imagined spaces carved in stone may have portrayed otherworldly territories, but at Chartres, they also taught and persuaded through their local placement and their engagement with social roles, economic activities, and devotional practices.

This is an instance where our representations of imagined spaces accrue rhetorical force in their particular location of display. The scale and social setting of the artistic program at the south porch augmented its authority for devotees, tradesmen, artisans, and pilgrims. This, in turn, reinforced the social and theological teachings conveyed in its central tympanum sculpture. That is, the tympanum sculpture represented authoritative space and was itself an object located in authoritative space. Its images affirmed the sympathetic link between the social space of the community and the overarching scheme of the cosmos.

In the case of the Last Judgment tympanum at Chartres, the visual depiction of apocalyptic space was grounded in a spatial framework with social, political, and economic dimensions. Here, otherworldly spaces were fraught with this-worldly implications, creating a negotiation of space between heaven and earth. We are invited to consider apocalyptic space as a projection toward the transcendent, or as an imagined place where the spatial order of our world is overturned, subverted or recast—but its narrative must rely on our own spatial experiences and frameworks. Even our most abstract and imagined scenarios, once depicted and emplaced, inherit social and political weight. This is true even for the most hypothetical maps of our last places.

Notes

1. Leonard L. Thompson, "Mapping an Apocalyptic World," *Sacred Places and Profane Spaces: Essays on the Geographics of Judaism, Christianity, and Islam*, Jamie Scott and Paul Simpson-Housley, eds. (Westport, CT: Greenwood Press, 1991), 115.
2. Kirsti Copeland, "Thinking with Oceans: Muthos, Revelation, and the Apocalypse of Paul," *The Visio Pauli and the Gnostic Apocalypse of Paul*, Jan N. Bremmer and István Czachesz, eds. (Leuven: Peeters, 2007), 77–104; Kelley Coblentz Bautch, *A Study of the Geography of 1 Enoch 17–19: "No One Has Seen What I Have Seen"* (Leiden: Brill, 2003); James Scott, "On Earth as in Heaven: The Apocalyptic Vision of World Geography from *Urzeit* to *Endzeit* According to the Book of Jubilees," *Geography and Ethnography: Perceptions of the World in Pre-Modern Societies*, Kurt A. Raaflaub and Richard J.A. Talbert, eds. (Chichester, UK: Wiley-Blackwell, 2010), 182–96.
3. Ellen Bradshaw Aitken, "The Landscape of Promise in the Apocalypse of Paul," *Walk in the Ways of Wisdom: Essays in Honor of Elisabeth Schüssler Fiorenza*, Shelly Matthews, Cynthia Briggs Kittredge, and Melanie Johnston-DeBaufre, eds. (Harrisburg, PA: Continuum International, 2003), 153–65; Kathryn M. Lopez, "Standing before the Throne of God: Critical Spatiality in Apocalyptic Scenes of Judgment," *Constructions of Space II: The Biblical City and Other Imagined Spaces,* Jon L. Berquist and Claudia V. Camp, eds. (New York: T&T Clark, 2008), 139–55; Tina Pippen, "The Ideology of Apocalyptic Space," *Constructions of Space II*, 156–70.
4. Martha Himmelfarb, *Ascent to Heaven in Jewish and Christian Apocalypses* (New York: Oxford University Press, 1993), 4.
5. For my analysis, I draw upon the work of Henri Lefebvre and his spatial triad, comprising three interconnected aspects: spaces of representation (conceptual or textual space), representations of space (social space that is organized and lived through images, metaphors, and symbols), and spatial practice (the daily and sensory reality of space, which can be examined according to movements of individuals within it). Particularly helpful is Lefebvre's attempt to connect mentally constructed spaces—our blueprints, our cosmologies, and our urban plans—to our built, lived, and social environment. Henri Lefebvre, *The Production of Space*, Donald Nicholson-Smith, trans. (Malden, MA: Blackwell, 1991), 14 and 38–39.
6. Peter Kurmann and Brigitte Kurmann-Schwartz, "Chartres Cathedral as a Work of Artistic Integration: Methodological Reflections," *Artistic Integration in Gothic Buildings*, Virginia Chieffo Raguin,

Kathryn Brush, and Peter Draper, eds. (Toronto: University of Toronto Press, 1995), 138.

7. Arthur Kingsley Porter, *Medieval Architecture: Its Origins and Development, with Lists of Monuments and Bibliographies*, Volume 2 (New York: Baker and Taylor, 1909), 135.
8. Emile Mâle, *The Gothic Image: Religious Art in France of the Thirteenth Century* (Boulder, CO: Westview, 1972), 365–67.
9. Moshe Barasch, "Apocalyptic Space," *Apocalyptic Time*, Albert I. Baumgarten, ed. (Leiden: Brill, 2000), 306.
10. Barasch, "Apocalyptic Space," 317–18.
11. See Virginia Chieffo Raguin and Sarah Stanbury, *Women's Space: Patronage, Place, and Gender in the Medieval Church* (Albany, NY: SUNY Press, 2005), 213–16; Jacques Le Goff, *The Medieval Imagination*, Arthur Goldhammer, trans. (Chicago: University of Chicago Press, 1992), 85.
12. Barasch, "Apocalyptic Space," 319.
13. Barasch, "Apocalyptic Space," 320.
14. Susan Bratton, *Environmental Values in Christian Art* (Albany, NY: State University of New York Press, 2008), 112.
15. Debra Hassig, "The Iconography of Rejection: Jews and Other Monstrous Races," *Image and Belief: Studies in Celebration of the Eightieth Anniversary of the Index of Christian Art*, Colum Hourihane, ed. (Princeton, NJ: Princeton University Press, 1999), 29.
16. Michael Camille, *Image on the Edge: The Margins of Medieval Art* (Cambridge, MA: Harvard University Press, 1992), 77. See also James Bugslag, "Ideology and Iconography in Chartres Cathedral: Jean Clement and the Oriflamme," *Zeitschrift für Kunstgeschichte* 61.4 (1998): 491.
17. Keith D. Lilley, "Cities of God? Medieval Urban Forms and Their Christian Symbolism," *Transactions of the Institute of British Geographers* 29.3 (2004): 296–313.
18. Jane Welch Williams, *Bread, Wine and Money: The Windows of the Trades at Chartres Cathedral* (Chicago: University of Chicago Press, 1993), 34–35; Kurmann and Kurmann-Schwartz, "Chartres Cathedral," 140.
19. Kurmann and Kurmann-Schwartz, "Chartres Cathedral," 142.
20. Vibeke Olson, "Movement, Metaphor and Memory: The Interactions between Pilgrims and Portal Programs," *Push Me, Pull You: Physical and Spatial Interaction in Late Medieval and Renaissance Art*, Vol. 2, Sarah Blick and Laura D. Gelfand, eds. (Leiden: Brill, 2011), 512.
21. Annette Lermack, "Spiritual Pilgrimage in the *Psalter of Bonne of Luxembourg*," *The Art, Science, and Technology of Medieval Travel*, Robert Bork and Andrea Kann, eds. (Burlington, VT: Ashgate, 2008), 109. Lermack uses the tympanum sculpture above the west portal at Chartres as her example.

22. William Dunning, *Changing Images of Pictorial Space: A History of Spatial Illusion in Painting* (Syracuse: Syracuse University Press, 1991), 11–12; Herbert L. Kessler, *Seeing Medieval Art* (Toronto: University of Toronto Press, 2004), 116.
23. Williams, *Bread, Wine and Money*, 51.
24. Laura Hollengreen, "From Medieval Sacred Place to Modern Secular Space: Changing Perspectives on the Cathedral and Town of Chartres," *Architecture as Experience: Radical Change in Spatial Practice*, Dana Arnold and Andrew Ballantyne, eds. (New York: Routledge, 2004), 91.
25. Simon Coleman and John Elsner, *Pilgrimage: Past and Present in the World Religions* (Cambridge, MA: Harvard University Press, 1995), 116.
26. James Bugslag, "Pilgrimage to Chartres: The Visual Evidence," *Art And Architecture Of Late Medieval Pilgrimage In Northern Europe And The British Isles*, Sarah Blick and Rita Tekippe, eds. (Leiden: Brill 2005), 156–59.
27. Bugslag, "Pilgrimage to Chartres," 183. See also Spitzer on the miracle folklore narratives associated with the Virgin that helped developed the cult at Chartres, including the 1262 collection by Jean le Marchant, *Miracles de Notre-Dame de Chartres*. Laura Spitzer, "The Cult of the Virgin and Gothic Sculpture: Evaluating Opposition in the Chartres West Facade Capital Frieze," *Gesta* 33.2 (1994): 141.
28. Spitzer, "Cult of the Virgin," 146.
29. Paul Williamson, *Gothic Sculpture, 1140–1300* (New Haven, CT: Yale University Press, 1998), 2. For general shifts in beliefs and images related to Last Judgment by the thirteenth century, see Willibald Sauerländer, *Gothic Sculpture in France, 1140–1270* (H. N. Abrams, 1973), 30; Paul Binski, *Medieval Death: Ritual and Representation* (Ithaca, NY: Cornell University Press, 1996), 24–25 and 208–9; Bratton, *Environmental Values*, 112–13; Caroline Walker Bynum, *The Resurrection of the Body in Western Christianity, 200–1336* (New York: Columbia University Press, 1995), 12–13.
30. Hollengreen, "From Medieval Sacred Place," 84. See also Spitzer, "Cult of the Virgin," 146 and Williams, *Bread, Wine and Money*, 39.
31. Hollengreen, "From Medieval Sacred Place," 105, n.17, observes that the last count to reside frequently in the castle was Thibaud V, who died in 1191. Thus, we have an instance of the urban, architectural symbol of the count's authority, even in the absence of his person.
32. Hollengreen, "From Medieval Sacred Place," 88; See also Williams, *Bread, Wine and Money*, 22.
33. Hollengreen, "From Medieval Sacred Place," 91.
34. Williams, *Bread, Wine and Money*, 39.
35. Whitney S. Stoddard, *Art and Architecture in Medieval France: Medieval Architecture, Sculpture, Stained Glass, Manuscripts, The*

Art of the Church Treasuries (Boulder, CO: Westview Press, 1972), 173–75.

36. Williams, *Bread, Wine and Money*, 135.
37. O. von Simpson, *The Gothic Cathedral: Origins of Gothic Architecture and the Medieval Concept of Order* (Princeton, NJ: Princeton University Press, 1988), 166–69. For a contrary view, see Barbara Abou-El-Haj, "Artistic Integration and Social Consensus," *Artistic Integration in Gothic Buildings*, 221.
38. Bugslag, "Pilgrimage to Chartres," 170. Coins of Chartres served as medieval pilgrimage tokens: issued at a mint claimed by Gauthier, the Bishop of Chartres, they were stamped with images of Mary's banner and veil, and became themselves associated with miracles. Williams, *Bread, Wine and Money*, 118–19.
39. Philip Sheldrake, *Spaces for the Sacred: Place, Memory, and Identity* (Baltimore, MD: Johns Hopkins University Press, 2001), 60; Coleman and Elsner, *Pilgrimage*, 115; cf. Kurmann and Kurmann-Schwartz, "Chartres Cathedral," 143 and Williams, *Bread, Wine and Money*, 20–25.
40. Kurmann and Kurmann-Schwartz, "Chartres Cathedral," 142; Williams, *Bread, Wine and Money*, 163, n. 32.

Chapter 8

Invisible Spaces between Cosmology and Biography: An Approach to the Afterlife in Visual Media

Daria Pezzoli-Olgiati

In many religious traditions, the spatial locations of the afterlife are vividly depicted in words and images.[1] Siduri's garden in the Akkadian *Epic of Gilgamesh*, the description of the heavenly Jerusalem in the *Book of Revelation* or in the *Fourth Book of Ezra*, the representation of Paradise on Persian carpets or Dante Alighieri's *Divina Commedia* are diverse illustrations of a recurrent paradox: in these works, remote or otherworldly spaces are portrayed as tangible and present, although they lie beyond human experience, and are inaccessible within the bounds of human life. This chapter deals with this peculiar tension in religious representations of the afterlife. It focuses upon collective ideas of worlds generated and mediated by religious symbol-systems that cannot be experienced in the present life. Afterlife imagery is a good example of religion's function in providing an orientation for human life by delineating spaces beyond the borders of what can be immediately perceived.[2] This aspect is particularly interesting for an approach to religion that is concerned with media and religious communication.[3]

Spatial images of the afterlife recur frequently in the history of European religion. They are attested by a broad range of texts, images, performances, and other media that can be found down the centuries from antiquity until today, as briefly recounted above.[4] It is indeed striking to note that in contemporary culture, images of the afterlife, of both positive and negative worlds after death, are widespread, particularly in visual media, film, art, and advertising.[5]

This chapter aims to discuss spatial approaches to religious imagery by considering visual and material aspects of religious orientations.

First, it defines the relationship between a general approach to "religion" and the spatial construction of the afterlife from a theoretical perspective. Second, it considers the relevance of visual communication in generating afterlife imagery. Third, select examples of visual representations of otherworlds beyond the boundaries of life are presented and analyzed, considering in particular the depiction of these otherworlds and their pragmatic dimensions. For this discussion, I have selected two examples from the fifteenth to the seventeenth centuries, focusing on Christian imagery in German-speaking regions. The results of the case studies are then synthesized and critically assessed by outlining the significance of the different levels of spatial afterlife represented from both a cosmological and a biographical perspective.

Religious Orientation and the Spatial Construction of the Afterlife

When considering the question of the spatial construction of the afterlife in religious symbol-systems, it seems appropriate to focus on the processes of communication that represent, transmit, regulate, and contribute to identity construction within religious practices.[6] Viewed from this perspective, religion is perceived as an integral component of culture. It is seen as a complex symbol-system that provides a general existential orientation; it produces meanings that are comprehensive and value-laden within a group; it receives, transforms, negotiates, and transmits images of the world through multiple media.[7] Following Clifford Geertz, religious images of the world tend to function as both a model of the world and a model for the world, shaping and signifying it in a binding way.[8]

The images of the world generated by religious communication encompass dimensions beyond human reality. Religious images of the world widen the scope of potential experiences beyond the boundaries of life. They shape these boundaries, attributing to them a particular significance. With reference to our topic, religious images of the afterlife typically demarcate the boundaries between life and death or between this world and the otherworld very clearly, for example by describing the journey of the soul from the corpse to a place in an otherworld. Seen from this perspective, the afterlife can be interpreted as a coping strategy for dealing with death and contingency within a religious framework. Accordingly, two aspects are particularly relevant: in religious conceptions of the world, afterlife images are situated on both a cosmological and a biographical level. On a cosmological level, the focus lies on the question of the spatial dimensions of the otherworlds where the dead reside. These places are constitutive parts of the cosmos as a whole. On a biographical level, the afterlife coincides with the ultimate destination of personal human existence: here the chronological and the spatial dimension of afterlife are interwoven. From a spatial perspective this link is often realized with the idea of a journey, where temporal and spatial categories can easily be combined.

The Role of Visuality in Representing the Afterlife

I have so far suggested that religious symbol-systems shape worlds that are not directly perceptible and, therefore, provide a general, existential orientation that encompasses the afterlife on both a cosmological and a biographical level, or, to put it another way, religious representation makes the invisible visible. The wide diversity of "other" worlds of the afterlife can be visualized by religious groups and/or individuals through various reception processes; they can be diffused in the broader context of culture, they can be accepted or rejected or simply negotiated and transformed.

In the remainder of this contribution, I will focus on a particular form of visualization that is enabled by visual media, scrutinizing the function of such media that presupposes the involvement of gaze and sight. The capacity of receiving religious images of the world through seeing cannot be taken into consideration without also thinking about intermedial processes, particularly the interaction between images and texts. Religious communication is pleonastic and repetitive and is often articulated by an elaborate network of references.

However, it is particularly interesting to consider visual representations of otherworlds beyond the boundaries of life in a discussion about religion and space, since the concrete perception of images implies the simultaneous presence of the viewer and the work within the same spatial plane. Seeing is a process that is strongly related to space. Only in the encounter between image and viewer can the visual work be perceived and meaning generated.[9] By looking at a representation of the afterlife, the viewer can figure inaccessible places that belong to a cosmological image of the world. Furthermore, she or he can see the ultimate destination of her or his life.

These considerations about the afterlife and visual media provide a general frame within which visual representations of otherworlds can be analyzed, contextualized, and compared. To locate images of otherworlds within the history of religion in Europe in a comparative and diachronic perspective, it becomes necessary to consider several significant transformations and evolutions in the perception of physical, social, and symbolic space.[10] From an Aristotelian conception of space to modern space concepts, from societies imprinted with religious and theological concepts to secular, differentiated social settings, from polytheistic to monotheistic systems in interaction with different cultures in different times, the variability and dynamics are too complex to be put into a single scheme. Equally, there is a range of attitudes toward the necessity and viability of defining religious symbol-systems. What I am proposing here is not meant to exclude other possible approaches but rather aims at clarifying the standpoint I am arguing from.

The Afterlife from a Spatial Perspective in Select Examples

In the context of this discussion about afterlife worlds, their spatial dimension and significance for collective and individual identity, I have selected visual sources belonging to the material aspect of religion; all of them are objects of common use and embedded in individual devotional practices. To retain some coherence within this contribution, I have chosen two images from German-speaking Christian regions of the fifteenth to the seventeenth centuries.

The first example is a retable from Regensburg (Germany) made in 1488. The second is an altar image where motifs from both the Late Middle Ages and the Baroque era are combined. This image, dated to 1677, was commissioned for an ossuary in Oberägeri (Switzerland, Canton Zug). The sources from both Regensburg and Oberägeri depict afterlife places. They reproduce similar, but not identical, cosmological concepts and relate them strongly to the life of their viewers.

The retable with bas-relief from Regensburg (figure 8.1) was commissioned by Elisabeth Englmayr (d. 1491) and her already deceased husband Sigmund Graner (d. 1484) for the collegiate church *Zur Alten Kapelle* in Regensburg. It contains a peculiar, late medieval representation of afterlife places. The composition is elaborate and merits a more detailed analysis. Nevertheless, in the context of this article, the accent lies on the representation of afterlife spaces and their relationship to the world of living people.

The retable may be structurally divided into two components: a large central image surrounded by eight smaller scenes.[11] In the upper register of the central scene Christ is depicted as the judge flanked by Mary and John. The traces of two angels are still recognizable to

Figure 8.1 Retable with bas-relief, 124 × 188 cm, 1488. On the lower edge: *herr Sigmund graner Elizabeta sein hausfrw m cccc lxxxviii.*

Source: Courtesy of Museen der Stadt Regensburg, HV 1415.

the right and left of Christ. In the lower register the donor family is presented. On the right hand side (from the viewer's perspective) there is Elisabeth Englmayr with her two daughters and the coat of arms, Saint Bartholomew, and Mary protecting both clergy (left) and lay people (right) within her garment. On the left hand side there is a depiction of her husband Sigmund Graner (although his coat of arms is missing) with Saint Simon and different medieval motifs of Christ: Christ, the winepress, the poor souls of Purgatory, and Christ showing his healing wound (*fons pietatis*).

On the upper parts of both wings of this altar screen, there are four scenes depicting a good-works series: the sacrifice of the mass is represented during the elevation of the host, a figure offers clothes to naked people (following the *Gospel of Matthew* 25:36). On the same level in the right wing, there is a further reference to the same text by showing a man welcoming a pilgrim into his house. The fourth small scene presents two alternative ways of addressing Christ in prayer: the clergy man on the left is deeply absorbed in the prayer; the lay man on the right is superficially addressing Christ, while, in fact, he is thinking of the different activities required for the smooth running of his household. For the interpretation of this scene a comparison with a woodcut from the middle of the fifteenth century is useful

Figure 8.2 Painted xylography, fifteenth century.

Source: Courtesy of Bayerische Staatsbibliothek, München, Clm 12714, inside cover.

(figure 8.2).[12] In the Regensburg retable the vices of avarice (a chest and a coffer), vanity (a women looking at a mirror), lust (a couple in a bed), gluttony (a table), and envy (a store room with barrels and a figure with a horse) are represented (from left above).

In all four scenes an angel takes a central object of the depiction (the host, a cloth, the pilgrim's bag, and a prayer lace)[13] and brings it to the lower register, in the second scene from the left. This last series of images is central to the question that we are investigating in this chapter. The retable makes a series of places visible that are beyond the human experiences represented above. The angels bringing the objects to the poor souls in Purgatory are the link between the different places of human existence during life and after death. In the left-hand lower scene, beneath the central panel, the four angels are portrayed as bringing the chalice, the prayer lace, the cloth, and the bag to the souls residing in Purgatory, which is depicted as a prison. The soul receiving the chalice is released as an angel opens the door. In the paired lower register scene on the right, beneath the central panel, there is an unusual and rare representation of Hell as a construction comprising four foundry furnaces with flaring flames. Hell is shut by a massive door with heavy chains; a devil and an angel control it. The interpretation of the two constructions on the lowest register of the outer wings is difficult to tease out. Philipp M. Halm proposes that they be interpreted as a continuation of the good-works cycle echoing Mt 25:35–36 that surrounds them, illustrating the motifs of freeing prisoners (left) and burying the dead (right). In this case the scene would represent an ossuary where the remains of the dead are preserved.[14] A different meaning is proposed by Peter Jensen (1994) who reads the whole lower register as a representation of different parts of the underworld. Reading from left to right: the first lower register scene depicts a prior stage of Hell out of which Christ freed Adam, Eve, and the patriarchs, image two depicts Purgatory, image three Hell, before concluding in image four with Limbo. This second interpretation appears the more convincing of the two proposals since the formal compositional features of the depicted scenes—the elements chosen, the colors, and the collocation within the whole retable—belongs rather to the representations of Purgatory and Hell.[15]

Considered from a spatial perspective, the retable offers a panorama of some central cosmological sites according to late medieval Christian belief. The eschatological place of judgment is at the same time the place where the donor and his family, the family saints, and Mary protect and intercede on behalf of these individuals: this central scene links the future of salvation and the present of piety and prayer. Time is here considered not as a chronological, linear development but as a dimension of Christian life where eschatological salvation is not only a goal but also a quality of human life. The scenes dedicated to the good works of Christian piety (contrasted in the prayer scene with a negative example) represent spaces of action where the living person can relate to the poor souls in Purgatory. They represent spaces within a Christian life where the concern for salvation is always present. With positive deeds one can help the poor souls in another dimension crossing, in a religious sense, the cosmological boundaries. The angels assume the role of intermediates by bringing the symbols of a pious life to Purgatory. The relationship between the living humans and the poor souls is depicted on a spatial level: the image illustrates different moments in the history of salvation and in Christian existence by using a very articulated topography of places on both this side and on the other side of death.

By looking at this image, the viewer can locate him- or herself within this representation of a Christian universe. The visualization of this world and its afterlife spaces has a strong pedagogical trait and interacts with other media, particularly with blockbooks and other printing.[16] By providing an encompassing image of the world the retable offers both a Christian topography and a normative orientation for its viewers.

Therefore, spatiality plays a major role not only within the religious representation mediated by the image but also in the relationship between viewer and the retable. The reception process is realized in space, in this case in the church, where the believer faces his or her life within the depicted universe. The representation of the donors can serve as identification figures for the viewer. They can be seen as an example of piety, since, in the act of offering the retable, they themselves perform a good work. In this context, it is interesting to outline that the altar screen was commissioned by Elisabeth Englmayr after the death of her husband. By providing a visualization of afterlife places and their relationship to life here and now, the retable with bas-relief reiterates a tradition and engages the audience in processes of orientation within the general concept of salvation. In this topography of afterlife, cosmological and biographical moments are merged together.

The second example was made for an ossuary in Oberägeri, a village in the Swiss Canton of Zug (see figure 8.3). Although it belongs to a later age, it can be seen as a Baroque actualization that stands in a line of continuity with late medieval topics, as the striking similarities with a xylography dated ca.1488 shows (figure 8.4). As was the case with the first image, figure 8.1 also illustrates a Christian cosmological concept, where both a spatial and a moral orientation are provided.

Differing from the Regensburg retable, however, this painting titled *SPECVLVM RATIONIS. Viue ut aeternum Viuas* is configured in the form of a mirror that is structurally divided into different spaces. The focus is not on the relationship between living people and the poor souls in Purgatory, but rather on an individual's life illustrated as a dangerous path across a bridge (central oval) while the main cosmological sites, according to a Christian tradition that was reemphasized in Catholicism after the Reformation, are situated at the outer edges of the mirror. Here four representations of places are intercalated with four sections containing angels with banners.[17] Optically, heaven, the place of eternal beatitude (*Æterna felicitas*) where Jesus Christ, God and the Holy Spirit and Mary reside, is opposed to the lowest place, where Purgatory and Hell (*Æterna Damnatio)* are depicted together, the latter contained in the jaws of a monster. This opposition, building a vertical axis, visually bisects a horizontal axis where the place of earthly life (*Vita Mundi*) contrasts with the place of judgment. Earthly life is presented in two possible variants. In the foreground five men

Figure 8.3 Oil painting on wood, 112 × 86 cm, 1677. On the lower banner: *R. D. IoannesYthTvginvs de Ægerin / Plebanvs in SarmenstorfArgovia svis / Impensis hanc Aram dieri fecit. M.D.C.LXXVII.*

Source: Courtesy of Bildarchiv Kantonale Denkmalpflege Zug.

are sitting at a dining table having an enjoyable time. In the background four men are working (first from left), hunting with a falcon, and praying. They are leading a virtuous life, while the people who are dining are considered exposed to the dangers of vice, since they drink, eat, and play cards. They are wearing expensive, rich, and colored dresses, while the virtuous men are in dark colors. On the other

Figure 8.4 Spiegel der Vernunft, painted xylography, ca.1488, 40.4 × 29.1 cm.
Source: Courtesy of Staatliche Graphische Sammlung, München, Inv. N. 118 319 D.

pole of the horizontal axis Christ is represented as the judge with sword and palm, his feet resting on the globe. The archangel Michael holds a balance. On the left a saint intercedes for the dead, while on the right a devil controls the entrance to Hell. This place is labeled *Extremū Iudiciū*, which indicates that the scene is understood as the final judgment. As we saw above with figure 8.1, the scene functions at the same time as a place of personal judgment: personal and cosmic levels merge together.

Distinguishing a vertical and a horizontal axis relating the four cardinal points to the outer edge is just one possible way of looking at this "mirror of reason." Alternatively, by focusing upon the speaking angels and their relationship to the four scenes already described, it is equally possible to read the eights parts in a continuous circular movement. Starting with the four men conducting a virtuous life in the western panel, a viewer's gaze relates them to the heavenly space. The banner immediately above stresses this link: *Considera beneficia Dei, et Vide ne per ingratitudinem Deum offendas* (Consider God's grace and to avoid offending God by ingratitude leads to heaven). A similar link is outlined by the banner of the angel on the top right section: *Considera ad quem finem creatus sis, serua mandata* (Consider the goal of life and keep the commandments, you will not fail in the judgment and the soul will be saved). The connection between the lower part of the judgment scene and Hell is also suggested visually and by means of the text: the entrance to Hell leads to jaws containing fire. The angel located just above and to the right of this image says: *Considera, quid moriturus iusto Iudici sis responsurus* (Consider that when you die you'll have to answer to the righteous Judge). A similar visual conjunction brings the gaze back to the starting point: *Considera unde corpus sumptum sit Et Inferni cruciatus aeternos* (Consider of what your body is made and the eternal tortures of Hell). Directing this banner to the five men sitting at the table, the angel recalls the vulnerability of the body and the possible effects of eternal punishments.

Whereas the outer edges of the mirror show four prominent places within a medieval Christian worldview, that are often revisited in Baroque times, the central part, the mirror in a strict sense, is fully dedicated to a human biography. A man is walking on a bridge without a parapet; he is a pilgrim praying *Tu scis D(omi)ne, sed miserere, fiat voluntas tua* (You know [everything] Lord, have mercy, your will be done). A devil tries to bring him back to the worldly, sinful life. The pilgrim is walking toward the summit of the bridge at which point stands a vertical construction combining a tree-trunk, a gothic pillar, and a crucifix. An angel points at two tablets (depicted as the Mosaic tablets) on which is written: *Diliges Dominum Deum et Proximum tuum sicut te ipsum* (Love the Lord [your] God and your neighbor as your self). The second half of the bridge leads to the place of Judgment. Death is waiting with bow and arrow poised. A banner reminds the reader *Mors nescia flecti* (death cannot be crushed). The clock outlines the uncertainty of the point in time when the pilgrim will be hit by the lethal arrow. Under the bridge, the inescapable end of human life is portrayed: the naked corpse of the pilgrim lies next to skulls, reiterating the motif of the transiency of life on the one hand and on the other, perhaps, evoking the stereotypical representation of Golgotha to which the crucifix always refers.

The inner part of the mirror relates a biographical with a cosmological program: the pilgrim travels the path of life coming from the *Vita Mundi* and heading to the place of *Extremū Iudiciū*. As already anticipated, the place of Judgment has a double connotation on both an individual and a cosmological level. It stresses the relevance of a virtuous individual life within the general worldview depicted in this composition.

The title *Speculum Rationis*, and also the rich employment of texts within the image, connect this work to the medieval tradition of *speculum* literature, an educational genre with a strong moral accent.[18] Therefore, on a pragmatic level, the oil painting can be seen as an image for meditation. As we have seen, the gaze of the viewer can follow different paths. The parallel, noted above, of the painted xylography from the fifteenth century itself stresses this usage of the image as a visualization for individual insight.[19] Although there are differences in the iconographical elements and, obviously, in the technique, both representations address an individual. The older document is written in German, the newer in Latin (showing the better level of education of the commissioner, a priest depicted in the lower left-hand corner, aged 56 years), but both address the viewer and reader in the singular.

With regard to the second work of art, it is also striking to note that the relevance of spatiality is not restricted to the representation of an individual Christian life and the history of salvation in the form of a cosmology. The encounter of the painting and the viewer in space generate the gaze as a form of meditation and education. The image visualizes the world in a different manner than how a Christian devotee sees it in everyday life; the mirror is a place of revelation, a means of looking beyond reality. The depiction of worldly life (*Vita Mundi*) can be interpreted as an identification moment between the receiver and the painting. But the image inserts human life into the broader context of salvation and, hence, generates an encompassing orientation in both the depicted space and the lived experience of the devotee. This is the function of the "mirror of reason;" to enhance a process of self-awareness and belonging as a Christian believer.[20]

Spatial Dimension and Religious Orientation

In the context of this volume, the investigation of the chosen visual sources focused on two aspects: spatial dimension and religious orientation. First, the analysis considered the spatial representation of each image, with particular attention paid to the relationship between worldly and afterlife places on the one hand and the relationship between biographical and cosmological spaces on the other. Second, the analysis paid attention to the function of the images as a part of a general orientation for the believers. Both examples are works of art that are part of the interior decoration of sacred buildings and, in this sense, are not movable. Nevertheless, both examples are linked to printing presses that were circulating, thanks to the technological innovations of the late medieval era. This aspect is crucial for the reception processes of images and the diffusion of knowledge necessary to approach the considered visualizations of places on this side and the other side of death. The reception of these images shows a clear, linear composition meant for education and meditation as part of a process of identification mediated by a Christian cosmology.

The iconographical depictions of afterlife topography show basic similarities in each image as they both essentially belong to the same late medieval tradition that can be traced for several centuries, as the second example demonstrates. The selected images share a common conception that I would designate as cultural imagery about the afterlife. Nevertheless, differences in the articulation of the various spatial locations of the afterlife can be easily recognized. The link between cosmology and the individual life on earth is realized by means of different strategies. For instance, elements like the Final Judgment, the representation of an individual's death or of a bridge between life and afterlife play a central role in linking general worldviews with an individual biography. The material and pragmatic setting of the images is also illuminating: the images are embedded in a religious practice in which values and moral concepts are transmitted.

Physical, Social, and Symbolic Space in the Light of Afterlife Conceptions

The analysis of the selected examples emphasizes different aspects of the represented space in visual media in this particular religious-historical setting. The depictions are embedded in precise contexts of devotion. They expect their viewers to relate a general cosmological conception of the world that includes spaces beyond the present world with their own biography and the path they are taking in life. These objects that link cosmology and personal salvation (or damnation) can be interpreted as a representation of different aspects of space.[21]

First, the depictions make a visual statement about the physical "reality" of afterlife space. Afterlife worlds are places that can be isolated from their iconographic context and considered on their own, whether one concentrates on the burning topography of Hell or on the sacred space in the presence of the trinity. Within the selected depictions, these places evidence physical characteristics of space by means of a certain light or darkness, the means of access, their architectural structure, and so on. They are represented as a physical extension of the present life, spaces that are equally as important and vivid as earthly places. Considered at this level, the religious function of representing invisible worlds can be understood as a visualization of physical space by means of visual media.

Second, the depiction of the afterlife places marks out a space of action. Therefore, it assumes a social, collective relevance. The visualization of otherworlds that human beings will reach after death is deeply embedded in everyday life as a dimension of human behavior and action. These images aim at underlining the significance of values and of a transcendent world not only for the sake of the individual, but also for the entire community. This aspect can be deduced not only from the depictions themselves but also from their pragmatic context. They are accessible to a broad audience, reproduce common ideas and concepts, and relate to multiple media and practices where the immediate relevance of this otherworldly knowledge is stressed. The depiction of the afterlife influences the social space where the believers are embedded.

Third, by making invisible places visible, afterlife space assumes a strong symbolic connotation. This religious representation provides a general conception of the world where cosmological and individual aspects are interwoven. It attributes to the invisible world beyond the boundaries of human life a precise meaning. The symbolic quality of space relates the physical quality of afterlife worlds with their social relevance and, therefore, gives to the whole system consistency and coherence. Religious symbol-systems tend to monopolize the interpretation of an afterlife space by linking its cosmological with its biographical relevance.

It is interesting to remember that this function of religious tradition is itself performed in space, at least in the case of visual media. The multilayered qualities of the afterlife are transmitted in the encounter between object and viewer. By looking at the image the believers access the invisible world, as they become visible in their physical, social, and symbolic relevance.

Notes

1. The English term "afterlife" lays particular stress on the temporal dimension, describing something that happens once life has ended. By contrast, the German "Jenseits," French "au-delà," or Italian "aldilà" focus on the spatial dimension of what occurs after death, indicating a space that is beyond the limits of life. In the context of this study, it could be argued that a more appropriate term may be "other world," which more explicitly stresses the spatial dimension, but as a non-native speaker, I prefer to retain the more common usage. For the question of terminology to be used in a comparative study of afterlife imagery, cf. Daria Pezzoli-Olgiati, "Approaching Afterlife Imagery: A Contemporary Glance at Ancient Concepts of Otherworldly Dimensions," in *Other Worlds and Their Relation to This World: Early Jewish and Ancient Christian Traditions,* Tobias Niklas et al., eds. (Leiden: Brill 2010), 1–15.
2. Cf. Fritz Stolz, *Grundzüge der Religionswissenschaft,* 3 Auflage (Göttingen: Vandenhoeck & Ruprecht, 2001), 94–100, and Fritz Stolz, "Paradise, Religionsgeschichtlich," in *TheologischeRealenzyklopädie*, Volume 25 (Berlin: Walter De Gruyter Inc, 1995), 705–8.
3. Cf. David Morgan, *The Sacred Gaze. Religious Visual Culture in Theory and Practice* (Los Angeles: University of California Press, 2005); Jamal Malik, Jörg Rüpke, and Theresa Wobbe, eds., *Religion und Medien. Vom Kultbild zum Internet* (Münster: Aschendorff, 2007); Birgit Meyer, ed., *Aesthetic Formations. Media, Religion, and the Senses* (New York: Palgrave, 2009); Bärbel Beinhauer-Köhler, Daria Pezzoli-Olgiati, and Joachim Valentin, eds., *ReligiöseBlicke—Blicke auf das Religiöse. Visualität und Religion* (Zürich: TheologischerVerlag, 2010); David Morgan, ed., *Religion and Material Culture. The Matter of Belief* (New York: Routledge, 2010).
4. For general overviews cf. Georges Minois, *Historie des enfers* (Paris: Fayard, 1991); Alan E. Bernstein, *The Formation of Hell: Death and Retribution in the Ancient and Early Christian Worlds* (Ithaca/London: Cornell University Press, 1993); Herbert Vorgrimler, *Geschichte der Hölle* (München: Wilhelm Fink, 1994); Christoph Auffarth, *Irdische Wege und himmlischer Lohn. Kreuzzug, Jerusalem und Fegefeuer in religionswissenscahftlicher Perspektive* (Göttingen: Vandenhoeck and Ruprecht, 2002); Alan F. Segal, *Life after Death: A History of the Afterlife in Western Religion* (New York: Doubleday, 2004).
5. Prominent recent works include: *Hereafter,* directed by Clint Eastwood (Burbank, CA: Warner, 2010); *The Tree of Life,* directed by Terrence Malik (Los Angeles, CA: Fox Searchlight, 2011); *Melancholia,* directed by Lars von Trier (Hvidover, Denmark: Zentropa, 2011).

An impressive work of contemporary art dealing with this topic is Tea Mäkipä, *World of Plenty*, Digital Pint 400×2670 cm, 2005 (World Expo exhibition in Aichi, Tea). In the domain of public advertising the Nespresso campaign series with George Clooney and John Malkovich also provide a good illustration of this tendency.

6. Cf. Stuart Hall, "The Work of Representation," in *Representation: Cultural Representations and Signifying Practices,* Stuart Hall, ed. (London: Sage, 1997), 15–64.
7. Cf. Stolz, *Grundzüge*; Daria Pezzoli-Olgiati, "Eine illustrierte Annäherung an das Verhältnis von Medien und Religion," in Beihauer-Köhler et al., *ReligiöseBlicke*, 245–66.
8. Clifford Geertz, "Religion as a Cultural System," in *The Interpretation of Cultures. Selected Essays* (New York: Basic Books, 1973), 87–125. Cf. also Susanne Lanwerd, *Religionsästhetik. Studienzum Verhältnis von Symbol und Sinnlichkeit* (Würzburg: Königshausen and Neumann, 2002), 176–82; Jonathan Z. Smith, *Map Is Not Territory* (Chicago: University of Chicago Press, 1978), 291.
9. Cf. Hans Belting, "Der Ort der Bilder," in *Das Erbe der Bilder. Kunst und modern Medien in den Kulturen der Welt*, Hans Belting and Lydia Haustein, eds. (München: Beck, 1998), 34–53; Gernot Böhme, "Der Raumleiblicher Anwesenheit und der Raumals Medium der Darstellung," in *Performativität und Medialität* (München: Wilhelm Fink, 2004), 129–40.
10. A convincing introduction is given by Markus Schroer, *Räume, Orte, Grenzen. Auf dem Weg zu einer Soziologie des Raums* (Frankfurt: Suhrkamp, 2006), 29–46.
11. Cf. Philipp M. Halm, "Ikonographische Studien zum Armen Seelen-Kultus. Mit 15 Abbildungen im Text," *Münchener Jahrbuch der bildenden Kunst* 12 (1921–22): 1–24; Peter Jezler, ed., *Hillem, Hölle, Fegefeuer. Das Jenseitsim Mittelalter* (Zürich: Neue Zürcher Zeitung, 1994), 190–91.
12. Philipp M. Halm, "Ikonographische Studien," 18. The indication of the source seems incorrect (figure 8.2).
13. The angel in the prayer scene is missed, the object is depicted in the lower register, in the second scene from the left (third angel).
14. Philipp M. Halm, "Ikonographische Studien," 18.
15. Peter Jezler, *Himmel, Hölle, Fegefeuer*, 190. Furthermore, the absence of any acting figures and/or of angels is an argument in favor of Jensen's interpretation.
16. Cf. Philip M. Halm, "Ikonographische Studien," 23. And the iconographical parallels printed in it.
17. Cf. Regula Odermatt-Bürgi,"'SpeculumRationis' aus dem Beinhaus von Oberägeri," in *Unterwegs. Religion in Kunst und Brauchtum*, Christine Kamm-Kyburz, Eugen Holz, Richard Kern, eds. (Zug: Die Kunstgesellschaft, 1984), without pages; Peter Jezler, *Himmel, Hölle, Fegefeuer*, 316–17.

18. An interesting introduction is provided by Einar Már Jónsson, *Le miroir. Naissance d'un genre littéraire* (Paris: Les belles lettres, 1995).
19. Cf. the description of the xylography and transcription of the texts in *Die Frühzeit des Holzschnitts*, exhibition catalog for "Staatliche Graphische Sammlung München," (25. Mai bis 23. August, München 1970; 23–34); W. L Schreiber, *Handbuch der Holz- und Metallschnitte des XV.Jahrhunderts*, Volume IV (Leipzig: Karl W. Hiersamann, 1927), 43–44.
20. On the reltionship between mirror and gaze cf. Hans Belting, *Florenz und Bagdad. Eine Westöstliche Geschichte des Blicks* (München: Beck, 2008), 282–83.
21. These categories of spaces are formulated following the contribution of Martina, Löw, "Die topologischen Dimensionender Kultur," in *Handbuch der Kulturwissenschaften: vol. 1: Grundlagen und Schlüsselbegriffe*, Friedrich Jaeger and Burkhard Liebsch, eds. (Stuttgart: Mezler, 2004), 46–59 and Löw, *Raumsoziologie* (Frankfurt a.M.: Suhrkamp 2001).

Section III

Representing Space in Contemporary Contexts

Chapter 9

Ornament and the Other: Sacred Spaces and Religious Identities in Andalusia*

S. Brent Plate

An hour's drive from the nearest sea, in a shopping mall in the center of Seville, Spain, is a popular restaurant and nightclub called "Buddha Del Mar." Its three-leveled interior conveys a chic modernism, augmented by South and Southeast Asian Buddhist imagery. Under the gaze of Siddhartha and a thumping bass, dancers drink and drinkers dance until the break of dawn.

Zooming out to the shopping mall in which Buddha Del Mar operates furthers the religio-cultural clash. The building is a converted train station originally constructed early in the twentieth century, styled in what was then a trendy architectural fashion: the highly ornamental Neo-Mudejar (figure 9.1). Neo-Mudejar revives the "original" Mudejar style that arose on the Iberian Peninsula beginning in the twelfth century, after the Christian *reconquista* (reconquest) of land that Muslims had ruled over for centuries. The Christians did not much tolerate Muslims (or Jews), but they appreciated their arts and crafts, and after some exiles and forced conversions, Mudejar emerged as a Christianized version of traditional Islamic ornamentation. So here we see a secular nightclub borrowing Buddhist imagery, set within an architectural style initially developed by Muslims, existing in a predominantly Roman Catholic Christian culture.

Not far from the religious-themed nightclub is the great Gothic cathedral of Seville, by volume one of the largest Christian churches in the world. The structure replaced the former central mosque of Seville, after the 1248 reconquest of the city by Ferdinand III. The mosque itself had been built over an older Visigothic church, which

Figure 9.1 "Plaza de Armas" shopping center, Sevilla, Spain. Neo-Mudejar style.
Source: Photo by S. Brent Plate.

was itself created on top of a Roman pagan temple. The building's most prominent feature, and Seville's most prominent icon, is "La Giralda," originally a giant minaret built by the ruling Almohads in the twelfth century, later converted to a bell tower by Christians. The massive tower is ornamented in geometric motifs that were common across the medieval Islamic empire.

From these scenarios, I begin with two questions that motivate my interests in this chapter. There are no straightforward answers to be given, but the possibilities serve as guides. First, what does it mean that the bell tower of the largest Gothic church in the world was primarily built by Muslims, for Muslim purposes, and is filled with ornamentation that reflects a Muslim worldview? And, second, why does a train station turned shopping center borrow an ornamental architectural style from a half millennium earlier, when all those who developed the art have long since left the very space in which it developed? What does a view of architectural space tell us about clashes and changes in economic, political, and religious identities within one geographic location?

In answering such questions, what becomes most critical is the interaction between viewers and buildings as ornament triggers ways of being in space and time. In the following, I argue that the use of ornament is utilized as a visual marking of identity, as the production of space becomes the production of identity. Meanwhile, changes in styles mirror new modes of identification, though through this reflexive process repressed elements of identity can also be revealed. Starting with an outline of the place of ornament within Islam, I note how Islamic ornamental styles developed specifically in Seville and greater Andalusia, and ultimately how they continued after Islam was no longer the dominant cultural-political-religious force, especially commenting here on the rise of the neo-Mudejar. Finally I chart three differing examples of the ongoing presence of Islamic-inspired ornamentation in modern European spaces, and how *orientalism* is bound to *ornamentalism*. Through a focus on ornament, we begin to see some interesting and at times odd connections between Christians and Muslims in the Medieval and Modern Iberian peninsula.

Whether nightclub or cathedral, mosque or temple, there is no preexistent abstract space, untethered from lived life, sacred or otherwise. Instead, "spaces are produced," as Henri Lefebvre has argued. People create space, while space creates people. For our interests here, spatial production occurs through social processes including the religious practices of those who enter and perform within the built environment, the use of symbolism by designers, commercial interests, and religio-political battles that aim for greater public visibility. If spaces are produced, they are also *reproduced*. "No space disappears in the course of growth and development," Lefebvre says further. "Social spaces interpenetrate one another and/or superimpose themselves upon one another."[1] In the spatial reproductions in Seville and beyond, cultures clash, identities are formed and reformed, power exerted, symbols made meaningful, memories invoked. Each new reproduced space, with new ornamental styles, in this particular geographic location, in the southernmost part of the southernmost European peninsula, is built up and out, superimposing, overtaking, and replacing; the old never fully erased, but palimpsestically showing through, reveals a history of growth and development, and a history of otherness: the forgotten, repressed, and defeated.

Mere Decoration? Ornament in Islamic Context

Ornament is "not generally essential to the structure of an object,"[2] and therefore a secondary and even expendable element of a work of art. In other words, ornamentation is a kind of marginalized other to the main body of an artwork or architectural site. Nonetheless, ornament can be deeply meaningful. For the purposes of this chapter, I would initially put it thus: ornament can alter the perceptual space of the object to which it is attached—vases shrink or enlarge while rooms distort depending on their surface decoration, and social identities are formed and reformed by the added imagery. The structure itself is altered by its decoration. Ornament reorients space, altering the perception of the object to which it is attached.

In his masterful work *The Mediation of Ornament*, Islamic art historian Oleg Grabar argues for the importance of ornament by suggesting, "There is a difference between 'filling' a space with a design and transforming an object by covering all or parts of its surfaces with that design." The latter can go so far as to "transform the very purpose of its carrier."[3] Grabar offers an expansive view, not arguing that ornament only operates one way and not another—filling with design *or* transforming an object—but he does reorient the conversation by thinking of ornament as an "intermediary" whose chief operation is to establish a relation between viewer and object. Not far from Marshall McLuhan's quip that the "medium is the message," Grabar shows how the ornament is the object. He takes a performative approach, asking after what it *does* instead of simply what it *is*, provocatively claiming "ornament itself can be the message that is communicated."[4]

I won't here elaborate on the many ways representational imagery functions in the Islamic arts as a whole, but with reference to ornament we can say that representation occurs through constricting the cosmically ordered natural world (flowers, stems, leaves, and trees, as well as geometric proportions and alphabetic characters) and reforming it as a medium that points beyond itself. By "artificially" representing the "natural," the observer is paradoxically submerged in true nature, the cosmos ordered by the singular Creator. We are in the realm of symbolism here, but a direct correspondence between object and meaning is not immediately apparent.

Following most critically from Grabar, what I am interested in are the various ways ornamental imagery is used, understood, and then reused and reunderstood. The point here is that meaning stems not simply from the object itself, nor from some apparent theology behind it, but from the interaction between object and observers. This interaction is part and parcel of the production of space.

Drawing on Space: Calligraphy and Arabesque

The term "ornamentation" encompasses various designs within the Islamic arts. Of interest here is the ornamentation found in/on architecture, including three discernible types of imagery: (1) calligraphy as epigraphy, (2) vegetal arabesques, and (3) geometrical arabesques. These have a fair amount of overlap, and often are simply seen together. All are related in their uses of repetition, production of rhythmic space, two-dimensionality, interlacing, and the utter and the often complete coverage of building walls—some have pejoratively called this the *horror vacui* (fear of empty space), but Ernst Gombrich positively called it *amor infiniti* (love of the infinite).[5]

The English term "calligraphy" comes from the Greek roots meaning "beautiful writing." In Arabic the term *khatt* is often used, and has connotations of a space "marked out" or an "outline."[6] It is also the term used in early Muslim urban planning, and key cities such as Basra, Kufa, and Fustat were built using principles of measurement related to variants of the term *khatta*. There is, then, already an architectural relation to writing within Islam.

Along with architecture, calligraphy is a key artistic contribution of Islam to world cultures. It is valued because it both venerates the Word and is aniconic. Its basic forms, the Arabic alphabet, are abstract, nonrepresentational figures that yet produce meaning in the eyes of its beholders. Islamic studies scholar Seyyed Hossein Nasr provides an explicit symbolic-theological definition of calligraphy that allows an initial understanding: "Islamic calligraphy is the visual embodiment of the crystallization of the spiritual realities contained in the Islamic revelation."[7] While Nasr clearly has Quranic calligraphy in mind, beautifully written words are used on all manner of surfaces in everyday life: on dishes and vases, metalwork, wall hangings, tiles, and etched into stucco on the walls of buildings. Sheila Blair suggests that the Quran "puts man's whole life under the sign of writing,"[8] and thus inscriptions are omnipresent. The earliest Islamic building, The Dome of the Rock (late seventh century CE), uses large bands of calligraphy, while new mosques built in the twenty-first century continue to implement wall writings.

Similarly, the arabesque is ornamentation consisting of stylized plant forms and strictly geometrical interlacing work. Titus Burckhardt suggests, in ways that shift straightforward definitions of representation, "In its more stylized versions, an arabesque in plant forms bears no more than a distant likeness to a plant. But it does represent a perfect transcription of the laws of rhythm into visual terms. Its unfolding is continuous, like a wave, with contrasting phases having various degrees of resonance."[9] The implicit symbolism and theological privileging here is important to note, though curiously Burckhardt himself gets a bit swept away as his objective description gives way to affect, as the ornament moves "like a wave," provoking "resonance." The symmetry and rhythm of the arabesque moves its viewers, immersing them in the ordered cosmos, as plant material evokes the paradisiacal "Garden of Allah." Here, space and time interpenetrate. "Strictly speaking, rhythm belongs not to space but to time, of which it is not the quantitative measure but the qualitative one. It is by the mediation of movement that rhythm is reestablished in the spatial dimension."[10] Ornament moves in space, triggering temporal rhythm.

Deeply implicated with the vegetal arabesque is its cousin, the geometric arabesque. Geometry, from ancient to present times, is a gateway to spiritual wisdom. Patterns are repeated, with interlocking stars and geometric shapes heading off in multiple angles covering vast surfaces. Here, the doctrine of *tawhid* is deeply invoked, as color and line intersections suggest a harmonious, cosmically ordered universe. Geometry is the grid, the divine yardstick that allows measurement of the structure of the natural world, and thus through knowledge of geometry, one can know Allah. More immediately, there is a story told of an Ottoman sultan from the sixteenth century who was transfixed before a geometrically ornamented lectern. The *minbar* seemed to dance before him, swaying and shifting perspectives. Delighted, he exclaimed, "God! God! What are these beautiful forms? Like wine, they instantly caused me to lose my head."

Each type of Islamic ornament is chiefly two-dimensional—though the stone and stucco work can be a low bas-relief—and it is through formal design that imagery comes alive, impacting the eyes and bodies of viewers, creating a practice that is theological in orientation, but operates primarily within the perceptive body of observers, and theology itself can arise from this point. Ecstatically the walls move us, swaying, like waves, like dance, like inebriation. Space is made vibrant, not through intellection but through sensory connections. In this way, the practices of a place create meaningful space. As Michel de Certeau surmises in a theoretical manner applicable to Islamic ornament, space is "actuated by the ensemble of movements deployed within it."[11] Space is kinetic, and it is both the people that move as well as ornamented walls.

Ornament and Andalusia

Abd al-Rahman—the man responsible for establishing Islam in the Iberian Peninsula and the founder of the Emirate based in Cordoba—and later Islamic rulers felt "the need to use art to enforce a sense of place and identity."[12] The Islamic visual culture produced in Andalusia between the eighth and fifteenth centuries made use of the preexisting Hispano-Roman and Visigothic styles, but transformed them by bringing in the styles found in the Umayyad dynasty of Damascus, as well as North African "Moorish" aesthetics. One of the key design elements that emerged was stucco carving, as craftsmen fashioned superstructural ornamentation on architectural sites. Some of the greatest spatial creations during this time, and that continue to enchant a millennium later, include the Mosque of Cordoba, the Alhambra in Granada, La Giralda in Seville, and dozens of former mosques that have been converted to churches throughout Andalusia. Many other places, like the central Mosque of Seville, have been almost completely eliminated, or as is the case with the Alcazar in Seville, a mixture of historical layers remains. Calligraphy and arabesques heavily mark all the sites.

Somehow, here in this place, the strands come together around the experience of seeing—seeing as a bodily exercise moving well beyond the eyes. Writing on the Islamic arts of Andalusia, Jerrilynn Dodds and Daniel Walker suggest: "In religious building and decoration, each of these visual sources contributed to the development of ingenious and complex aniconic ornament that engaged the viewer in a contemplative rather than an empathetic relationship, which distinguishes Islamic religious arts from those of Christians."[13] Here is a succinct statement on the ways that religious and art historical approaches to the visual arts should not be simply rooted in struggles to determine symbolic value and iconographic reference, but need to be seen as imbedded within practices of adherents experiencing and creating space. In the Islamic vein, bypassing iconic representation, ornamentation can trigger a contemplative space.

Mudejar

The *reconquista*, finalized in 1492 with the fall of the last Muslim rulers in Granada, established a Spanish national identity based on the unification of Iberian peninsular provinces under a single monarchy, the consolidation of a common language (Castilian), and a common religion (Roman Catholicism). The establishment of the linguistic-religious-political confederation of Spain is simultaneously marked by the expulsion of Muslims and Jews from the Iberian Peninsula. This is the official story written by the winners using verbal language, yet if we look closer we find that this "other" can never be wholly eliminated and vestiges of a previous culture will always be in evidence, as it is in the ornamentation of monuments throughout Andalusia. There is a continued prevalence of Islamic ornamental style in Andalusia after the conversion, expulsion, and/or "domestication" of Muslims, a style that came to be known as *Mudejar*. As the Christians came to power, they set out on an extensive building and rebuilding campaign. It was critical to rebrand Moorish palaces as Spanish palaces, mosques as churches, and minarets as bell towers.

The term "Mudejar" was initially a pejorative term for those Moors who remained in Andalusia after the Christian *reconquista*.[14] In the nineteenth century, the Spanish art historian Amador de los Ríos used the term Mudejar for visual phenomena, and it has become particularly germane when discussing architecture.[15] Thus, to speak of Mudejar today, one is generally discussing an architectural ornamental style that emerged after the *reconquista*, when Christians were in power, while Moors ornamented the buildings.

The Mudejar arts utilize "poor" materials (e.g., brick instead of marble), while they were put to use in the most prominent of places: palaces, churches, and cathedrals. Kings, princes, and other leaders did not seek to lavish their spaces with fine and costly material, but with fine craftsmanship—vast and intricate ornament became a symbol of power. The Christian monarch Peter the Cruel's palace in Seville (the Alcazar, adjacent to the cathedral) transformed the earlier Almohad palace, and borrowed freely from styles of the Muslim-built Alhambra, with halls of plasterwork, brick, and stucco. Today one can find Arabic epigraphy that praises Allah on one wall, while other walls inscribe praise to Peter the Cruel. The mixture here is so complex that it is unclear what parts were built when, by whom, and for what purpose. Other places, like the already mentioned Giralda, and San Marcos's church towers in Seville, make the division easier: Almohad Muslims built the base as a minaret, and Christians affixed a bell tower to the top, remaking the meaning of the site. This is not merely visual, but points toward the role of *sound* in the construction of space. The term "minaret" stems from the Arabic *manara*, which means "a place that gives light," cousin to the Hebrew *menorah*. Sound and sight are intertwined, as space is sensually produced and reproduced. Such "additions" resonate a deep metaphoricity: Modern, Christian Spain is built upon the work on Medieval Muslims. It is seen here architecturally, but the same is true for philosophy, theology, and identity.

The collective studies of Jerrilyn Dodds, María Rosa Menocal, and Abigail Krasner Balbale have shown the hybridity that stands at the heart of contemporary Spanish identity. In their book *The Arts of Intimacy: Christians, Jews, and Muslims in the Making of Castilian Culture*, they claim:

> The written, verbal, and artistic languages that were shared by Muslims and Christians, the meanings attached to these forms and styles, became

> the cultural bedrock of the Castilians, and of the Spanish nation-state that followed. Castile had learned from its common space with Islam and Judaism how to create an identity through culture... *Perhaps Castilian culture did not develop along a path at all, but instead in a series of spaces—castles, cities, battlefields, courts, mosques, synagogues, and cloisters—spaces destroyed and redrawn scores of times over the centuries.* It was formed from the competition, dominion, envy, and assimilation that occurred in those places shared with different Christians, Muslims, and Jews. (Italics added)[16]

The very heart of modern Spanish identity, these authors argue, rests on the artistic creations across times, political structures, and religious traditions. The nation-state is created in and through the production and reproduction of spaces, and their forms and styles.

Mudejars and their styles inevitably disappeared, owing to changing artistic tastes as well as the real practicalities that all remaining Moors were permanently expelled from the peninsula through a series of decrees in the early seventeenth century. Their artwork lived on.

Ornamentalism and Orientalism: The Revival of Islamic and Mudejar Ornamentation in Modern Europe

While Muslims, at least until the 1990s, have been almost wholly missing from modern Spain, traces of Islamic culture remained scattered in the corners of the Spanish geographic psyche. One of the more vibrant trends in the Modern period in the South of Spain, mentioned at the opening of this chapter, are revivals of the Moorish and Mudejar styles, including the Neo-Mudejar. These styles now take on a quasi-nostalgic reminder of the glorious past of Andalusia. The final section of this chapter looks at three widely diverging buildings that begin with Moorish ornamental style, while adapting it to contemporary uses.

The Blas Infante House, Seville

One of the key figures in modern Andalusian identity was Blas Infante, born in 1885 and killed by Franco's troops in 1936, around the same time as his friend Federico Garcia Lorca's assassination. In 1915 he wrote *The Ideal Andalusia*, becoming known as the "father" of Andalusia. In writing about the place, he sought to reimagine a unified past history of Al-Andalus in order to create a different future for the region. His inclusive attempts took on visual form as well as verbal, reimagining as well as re-*imaging* a synthesized past (however "ideal") that would provide a synthetic basis for the future. Infante designed the symbolic motifs for the Andalusian flag, and built a small house/museum on the outskirts of Seville, known as La Casa Blas Infante.

For the House, Infante utilized the symbolic ideas of an esoteric religious movement that was popular in his time—namely, Theosophy—for the basis of the ornamental work in the architecture. In the House, however, the mystical-religious ideas of theosophy are subverted to a political end. Seeking to be both locally relevant and universally oriented, Blas Infante created his Casa with a strong concern to relate the local, indigenous Andalusian culture to a more universal understanding of humanity. I have suggested that Blas Infante attempted to create a "visual Esperanto" with his all-inclusive imagery.[17]

If the hybridity of Islamic ornamentation that developed in medieval Iberia pulled off a beautiful mixture of influences that coalesced yet retained boundaries, Infante's attempts at unification resulted in something a bit different. Infante chose an all-inclusive approach to pertinent symbols. Visually, la Casa is a jumbled conglomeration of artistic styles. The walls include tile work, plaster work, wood work, and painted decorations accomplished in styles ranging from Christian, Jewish, and Muslim sources, and from almost a millennium full of various modes. The visitor sees Renaissance styles next to Gothic styles, Islamic-inspired decoration next to Jewish symbols.

Even so, in a quest for universal symbols, some reduction is necessary. Typically, figural representation (animals, trees, humans) is out. Except for the coat of arms on the Andalusian flag, Infante's House is devoid of figural representation. Infante used abstraction, as did contemporary artists, but saw it as historically situated. In his House, the repeated star-motif on the doors is a key symbolic image that reveals a quest for a visual Esperanto that is both indigenous to Andalusia and is universal (figure 9.2). These doors attempt to rescue the Islamic and, secondarily, Jewish past of Andalusia, and reconcile it with the primarily Christian present. They are clearly an imitation of the Islamic ornamented doors seen throughout Andalusia as well as across Muslim lands from Iran to Jordan to Morocco. The stars—typically six-pointed and five-pointed—indicate a relation to a number of traditions and, as they do in theosophy, the stars become synergetic: up and down, right and left, are portrayed in creative tension with each other, getting close to the geometric arabesque. By comparing the ornamentation on the doors and walls of Infante's House with the Islamic and Mudejar styles, we see again how this is not

Figure 9.2 Interior doors and ornament in the Blas Infante House, Seville.

Source: Photo by S. Brent Plate.

mere decoration, but is intended to reveal something about deeper, cosmological structures of existence, connected to the political realities of the here and now. Through his House, Infante creates a visual mythology for an Ideal Andalusia. And because he borrows imagery from a specific Islamic and Jewish past that is rooted in Andalusia, he also points toward the promise of a more universal history that might be reimagined in the future.

The New Synagogue, Berlin

Significantly, Islamic ornamental styles, via the Mudejar, emerged again across Europe in new forms and new spaces. One of the most prominent uses was to be seen in synagogue architecture, and here I use one example among many. As part of the Jewish Reform movement after Emancipation across Europe, the New (Neue) Synagogue was founded on Oranienburger Strasse in Berlin. Dedicated on Rosh Hashanah, 1866, its architect was Eduard Knoblauch. With seating for over 3,000 people the building became a spatial symbol of a renewed Jewish life in Berlin, its large, gleaming dome visible all around, and sitting not far from Berlin's Jewish Quarter. It was often talked of as an "ornament" for the city of Berlin itself.

The "Moorish/Oriental" style was initially used for a few reasons in this synagogue, as it was across Europe and North America. For one, the Gothic, neo-Gothic, and Romanesque styles that were prominent in Christianized Europe at the time were not thought suitable for a Jewish place of worship. After all, these were "Christian" and "Western" architectural forms. To find their own style, Jewish communities turned to the history of Islamic architecture. As one newspaper article put it, the New Synagogue "leads us into the fantastic wonder of the modern Alhambra with all the thousand-fold magic of the Moorish style."[18] Meanwhile, architectural historian Dominique Jarrassé suggests, "The Jews had no architecture of their own, but their Semitic cousins the Arabs did; and not only was this heritage a remarkable one, it was also unambiguously different from that of Christian and the West in general."[19] Some even remarked that the New Synagogue looked like a mosque, which furthers a long-standing relation of identities between Muslims and Jews, especially as they were both lumped together and labeled "other" from the Modern European perspective.[20]

So, without any reigning plans about how a synagogue should be built, and with a fair tinge of Romanticism about the "Golden Age" of Jewish culture, Andalusian styles were borrowed for the Berlin building. Synagogues across Europe and North America from the mid-to-late nineteenth century were built in a style that harkened back to this time. In space, a time was reproduced. The older Moorish style synagogue thus became a visual symbol of new Jewish identity, and what was specifically "Moorish" was the ornamental styles: slender pillars, horseshoe arches, ribbed arches, epigraphical elements, and arabesque designs on the walls. The Hebrew writing over the entry mimics a calligraphic style of writing seen around entrances to mosques, mausoleums, and other spaces across the Islamic world. As with the visual-spatial power of arabesques, the Synagogue's ornament impacted its observers, as art historian Lothar Brieger recalled its "mixed effect of exaltation and losing oneself in a holy space."[21]

This was "oriental*ism*" and it had its detractors, especially when a renewed wave of anti-Semitism spread across Europe later in the nineteenth century. One Paul de Lagarde wrote critically of the New Synagogue, stating, "Through the style of their synagogue, the Jews emphasize their alien nature."[22] And Richard Wagner complained of the New Synagogue's "insolent luxury," calling it "pretentious."[23] The use of the Moorish ornamentation in urban centers emphasized the exotic, "othering" of Jews.

Yet, this was not simply an othering created by the Christians, and the orientalist impulses had its champions within the Jewish world as well. Jews were searching for a sense of identity in Christian-dominated cities after Enlightenment and Emancipation. For many central European Jews, the "Oriental" label was to be admired. Even into the beginnings of the twentieth century, in the midst of a spiritually desolate Europe, Jewish intellectuals like Martin Buber sought to distinguish Jews from the Christian Europeans. In 1916, after living in Berlin for many years, Buber's essay "The Spirit of the Orient and Judaism" was published. It is an interesting document, somewhat essentializing and Romantic, arguing that in spite of exile, "the Jew has remained an Oriental."[24] I note this here simply to indicate that "Orientalism" has historically had a number of connotations to it.

On this side of Nazi genocide, it is difficult to see the orientalist impulse as having been a good idea, but the example of the New Synagogue shows again the power of ornament in the broader constructions of identity and culture, via the reproduction of space. If Blas Infante wanted to show an all-inclusive past by incorporating multiple strands of historical ornament to show a plural culture, the designers of the New Synagogue used ornament to enable a distinctive spatial identity, set apart from the surrounding environs.

The Morabitun Mosque, Granada

Turning back to Andalusia and the ornamentalizing/Orientalizing projects, I finally take a brief look at the new Mosque of Granada. Opened in 2003, this is the first public mosque in Spain since 1492. Like the Muslim Park 51 project in Manhattan, and the New Synagogue in Berlin, it is dedicated to education, cultural events, as well as housing prayer. And, like Park 51, its construction was surrounded by struggle.

European converts to Islam formed the mosque, after meeting in various places and under various rubrics since the 1980s. Much of the language used by the group entails a "return" of Islam to Spain. Indeed, its location at the top of the Albaicín (the old "Moorish quarter" of Granada) looks directly across to the Alhambra, and in a statement the community wished it to be "a clear and unmistakable symbol of the return of Islam to this land."[25] The architectural influences are an amalgam of Turkish, Arabian, North African, and Andalusian elements. The group calls itself the Morabitun (after the Almoravid Kingdom that ruled the Maghreb, including half of the Iberian peninsula between the eleventh and twelfth centuries). Many groups, especially from the dominant Catholic communities in the area, protested the building of the mosque, but its construction was allowed to go ahead as long as several stipulations were met, including making the garden accessible to the public in the middle of the day, and having the exterior assimilate with the architectural environs. Moorish and Mudejar ornamental styles emerge at times, though the clean white-painted brick walls keep it from being overbearing, blending in with the white walls seen across the Albaicín, and the tower looks more like a bell tower than a minaret, making it fit with some of the Mudejar churches in the area.

The interior is a bit different, with somewhat more ornamentation, including a *mihrab* that mimics that of the Great Mosque of Cordoba, and calligraphed tile work, but apart from a few particulars it too is marked by a striking *lack* of ornamentation, considering all that has been noted previously in this chapter. There is no sense of the *horror vacui/ amor infiniti* of the Alhambra or Cordoba mosque. The interior designer of the mosque, Sidi Karim, states, "We are not looking back nostalgically. I have built a 2003 mosque." In a study of the mosque, Oskar Kerkaaik follows Karim's statement by saying, "there is no discrepancy in calling a praying room that clearly and deliberately bears the traces of various historical Islamic monuments 'a 2003 mosque.' Imitation serves to create an atmosphere that is supportive of the timeless act of praying. The result is a 2003 interpretation of the infinite."[26]

Conclusions

My interests here have been twofold. First, I have attempted to demonstrate the power of ornament, an element often overlooked in studies of religion and the arts, as well as lacking in studies of sacred spaces. As I have argued, the production of religious space occurs, in part, through use and reuse of ornamentation. There is no "mere ornament"; rather, ornamental imagery creates architectural environments, and it is in the effects of epigraphal and arabesque forms that space is produced and reproduced. My second aim has been to relate ornament to otherness and social identity. Analogically, we see that just as ornament is external to the work of art, so is ethnic and religious foreignness outside the establishment of national and cultural identities. Yet, in the end, there can be no final separation of what is "merely ornamental" from the essential work of art, for the ornamental alters the structure on which it exists, just as the "other" alters the "sameness" of identity.

Ornament's affects often bypass symbolic analysis. Vegetal arabesques may explicitly re-present the "Garden of Allah," or implicitly re-present an ordered cosmos, though just as often, ornamentation does *not* point toward transcendence, but toward immanence, co-constituting a space in which people are moved "like waves." What makes spaces meaningful is not simply the structure itself, but the interactions with and within the space, generated by the intermediaries of ornament. "Mere ornament" is a misnomer. Walls can talk, moving people to action, to violence, to contemplation, to ecstasy, ultimately fulfilling a need for an identity formed visually, sonically, and otherwise sensually.

Notes

* Some of the research for this chapter was undertaken through grants from the American Academy of Religion's Southeastern Region, and the Association for Religion and Intellectual Life. Special thanks are due to Alberto Egea of the Centro de Estudios Andaluces, and Edna Rodríguez-Plate for assistance in research.

1. Henri Lefebvre, *The Production of Space* (Oxford, UK: Blackwell, 1991), 86.
2. Dominique Collon et al., *Grove Art Online*, s.v. "Ornament and Pattern," accessed June 28, 2013, http://www.oxfordartonline.com/subscriber/article/grove/art/T063880.
3. Oleg Grabar, *The Mediation of Ornament* (Princeton, NJ: Princeton University Press, 1992), 41.
4. Grabar, *Meditation*, 44. Important precursors to Grabar's study include E. H. Gombrich, *The Sense of Order: A Study in the Psychology of Decorative Art*, 2nd ed. (London: Phaidon Press, 1984).
5. See Gombrich, *The Sense of Order*.

6. See Grabar, *Mediation*, 66. See also Jamel Akbar, "Khat[.]t[.]a and the Territorial Structure of Early Muslim Towns," *Muqarnas* 6 (1989): 22–32.
7. Seyyed Hossein Nasr, "The Spiritual Message of Islamic Calligraphy," in *Religion, Art, and Visual Culture: A Cross-Cultural Reader*, S. Brent Plate, ed. (New York: Palgrave, 2002), 113. See also the definitive work, Sheila Blair, *Islamic Calligraphy* (Edinburgh: Edinburgh University Press, 2006).
8. Sheila Blair, *Islamic Inscriptions* (New York: New York University Press, 1998), 11.
9. Titus Burckhardt and Roland Michaud, *Art of Islam: Language and Meaning* (London: World of Islam Festival, 1976), 66.
10. Burckhardt and Michaud, *Art of Islam,* 66.
11. Michel de Certeau, *The Practice of Everyday Life* (Berkeley: University of California Press, 1984), 117.
12. Jerrilynn Denise Dodds, *Al-Andalus: The Art of Islamic Spain* (New York: Metropolitan Museum of Art: Distributed by H. N. Abrams, 1992), xxiii. For the full story of Abd al-Rahman and the establishment of Andalusia, see Maria Rosa Menocal, *The Ornament of the World: How Muslims, Jews, and Christians Created a Culture of Tolerance in Medieval Spain* (Boston, MA: Little, Brown, 2002).
13. Dodds, *Al-Andalus*, xx.
14. See the chapter "Mudejar Status," in L. P. Harvey, *Islamic Spain, 1250 to 1500* (Chicago: University of Chicago Press, 1990), 55–73.
15. José Amador de los Ríos, Pedro de Madrazo, *El estilo Mudejar en arquitectura: discurso* (Valencia: Librerías "Paris-Valencia," 1996 [1872]).
16. Jerrilynn Denise Dodds, Maria Rosa Menocal, and Abigail Krasner Balbale, *The Arts of Intimacy: Christians, Jews, and Muslims in the Making of Castilian Culture* (New Haven, CT: Yale University Press, 2008), 6.
17. S. Brent Plate, "El mundo de Infante a través de la simbología universal de las estrellas," in *La Casa de Blas Infante en Coria del Río*, Alberto Egea, ed. (Sevilla: Centro de Estudios Andaluces, 2005), 171–84.
18. From an 1866 article in the *National Zeitung*, quoted in Hermann Simon, *The New Synagogue, Berlin: Past-Present-Future*, 3rd ed. (Berlin: Edition Hentrich, 1999), 11.
19. Dominique Jarrassé, *Synagogues: Architecture and Jewish Identity* (Paris: Vilo, 2001), 178.
20. In a chapter called "Muslim Jews," Anouar Majid tells how Jewish and Muslim identities were identical categories in the European imagination. Anouar Majid, *We are all Moors: Ending Centuries of Crusades Against Muslims and Other Minorities* (Minneapolis: University of Minnesota Press, 2009).
21. Quoted in Simon, *The New Synagogue, Berlin*, 11.

22. Simon, *New Synagogue*, 10.
23. Quoted in Jarrassé, *Synagogues*, 180.
24. Martin Buber, "The Spirit of the Orient and Judaism," in *On Judaism* (New York: Schocken, 1996 [1916]), 56–78 (75).
25. From "La mezquita de Granada, una realidad," *El País Islamico* (June 1998), 16. Quoted in Jennifer Roberson, "Visions of Al-Andalus in Twentieth-Century Spanish Mosque Architecture," in *Revisiting Al-Andalus: The Medieval and Early Modern Iberian World*, Glaire D. Anderson and Mariam Rosser-Owen, eds. (Leiden: Brill, 2007), 247–71 (264).
26. Oskar Verkaaik, "In Search of the Urban Bedou: The New Mosque of Granada," unpublished manuscript.

Chapter 10

Spaces of Ridicule: Spatial Differentiation of the Caricature in Eighteenth-Century England from the Perspective of the Study of Religion

Monika Glavac

In modern times, caricatures are commonly found in newspapers, satirical magazines, illustrated books, or on the Internet. They are situated within different media and their positioning varies according to the particular type of media in which they are published. This is especially the case if they are situated in the virtual world of the Internet, where space and time take on another dimension. It is also true, however, that in our own time caricatures have tended to play a somewhat more marginal role, as they represent only one among the whole host of possibilities for leisure, amusement, or critique. Nonetheless, in recent years caricatures have once again begun to attract attention, especially in relation to the Mohammed cartoons controversy that was caused by the publication of twelve caricatures in the Danish newspaper *Jyllands-Posten*.[1] Such controversy is in keeping with the history of the European caricature, as caricatures were often involved in conflicts, particularly in nineteenth-century France, when they were constrained by censorship.[2]

Eighteenth-century England and nineteenth-century France are both widely regarded as golden eras in the history of the European caricature, eras in which the production and consumption of caricatures flourished and aroused the greatest public interest. In my article I focus on the former of these two eras, because it enables me to reflect on the specific subject of space on a range of different levels and in connection with different aspects. First, I discuss space and caricature by analyzing and reconstructing the geographical, economic, and sociospatial contexts. In eighteenth-century England caricatures were part of the print industry that was located at specific sites in the city of London. Caricature shops were established where caricatures were produced, sold, bought, or viewed. In this context, space is understood as a complex network of different aspects, notably the interplay of town, streets, buildings, and places of consumption. Additionally, people act in this space and by so doing they construct space and give it meaning.[3]

Second, the focus shifts to the representation of space within caricatures. The question of how space is represented involves a transition to more aesthetic reflections on spatiality. Space is thereby understood as a representation in an aesthetic object, or as Winfried Fluck writes, space is "imaginary space."[4] To exemplify these theoretical reflections I wish to analyze the engraving *Credulity, Superstition and Fanaticism. A Medley* (1762) by William Hogarth. Hogarth depicts a complex network of interior and exterior, above and below, and in so doing, generates meaning. Furthermore, works of art are addressed to a viewer,[5] and with a viewer is added the category of the inside (of the picture) and the outside (of the viewer). This category adds another dimension to space: the two-dimensionality of the representation interacts with the multidimensionality of its reception.

To summarize the main aim of this article is to understand the spatial differentiation of caricatures in eighteenth-century England and the representation of space in images that are connected with a viewer.

Space and Caricature: A Geographical, Economic, and Social Interplay

There has been little attempt to explore space as a theoretical term in connection with caricatures, although the relationship between the print industry, the location of caricature shops, shopkeepers, and consumers of the English caricature in the eighteenth century has attracted some interest.[6] In this section of the article, "space" refers to a network of geographical aspects, such as the location of the caricature shops, economic aspects such as the extent of the print industry in London of the eighteenth century, and social aspects such as the gender and class of the consumer. Space may therefore be defined as the interplay between geographical, economic, and social characteristics.[7]

The following—anonymous—illustration entitled *Caricature Shop* (figure 10.1) may act as a starting point to explore the relationship between space, as defined above, and caricature.[8]

Figure 10.1 Anonymous, Caricature Shop, 1801, etching.
Source: Courtesy of The Lewis Walpole Library, Yale University.

The caricature depicts a crowd standing in front of a shop who are viewing caricatures. The shop is located in the district of Holborn in north London, and it is believed that the figure on the left-hand side is P. Roberts, the shop owner himself. The audience includes people of different social ranks: young and old, black and white, rich and poor, and even a dog. Based on this picture, one may ask: Where were caricatures produced and distributed and who was in charge of their production and distribution? This raises further questions: Where were caricatures displayed and consumed, and who bought and/or viewed them? These questions concerning the production, distribution, and viewing of caricatures will guide the analysis that follows, which aims to offer an impression of the spatial differentiation of the caricature in eighteenth-century England.

Caricature shops were part of the growing print industry in London. Caricatures were traded mostly in established print shops located in the traditional center of printing and publishing situated around St. Paul's Cathedral and Fleet Street whose influence spread all the way to Westminster Palace. It is noteworthy that not only do economic and spatial aspects correspond with one other but also that economic aspects have a significant impact on the process of creating space. The space of the town becomes specialist: the print industry in the eighteenth century is located in specific streets and areas of London.

Another significant factor in the creation of space is society: people create space.[9] The location of the print and caricature shops corresponds, for example, to the social rank of the customers. Furthermore, the content and the quality of the caricatures varied depending upon the location of the shop. The print shops that were located in the West End sold high-quality caricatures with political content to upper-class customers, whereas the print shops located around St. Paul's Cathedral served the lower classes.

A famous print shop at the time was Hannah Humphrey's, located near Westminster Palace. She sold most of the caricatures produced by James Gillray, who is deemed to be the most well-known English caricaturist of the eighteenth century. His caricature *Very Slippy Weather* (figure 10.2) shows people standing in front of Hannah Humphrey's shop viewing his caricatures that are displayed in the shop window.

It is remarkable that political caricatures were sold in the direct vicinity of parliament. This was possible owing to relaxed censorship laws, which encouraged freedom of expression, and was promoted by the marked public interest in politics. As the two caricatures already referred to show, the windows of caricature shops were used to display caricatures in order to exhibit them to the people. Jon Stobart, Andrew Hann, and Victoria Morgan characterize the shops of eighteenth-century England as follows:

> It is the front window upon which much attention has focused. Displays set up there were designed to encourage people to enter the shop, with shopkeepers going to considerable trouble and expense to produce a pleasing spectacle. Thus, the shop was transformed into a place for display and marketing: it was a site of commodities, information and persuasion.[10]

Exhibitions also frequently took place inside caricature shops (see figure 10.3). The print or caricature shops, therefore, created correspondences between inside and outside: the product is presented to potential customers passing on the street as well as displayed and sold in the interior. By putting on exhibitions within the shop, the shop turns into a public social space in which not only the commercial roles

Figure 10.2 James Gillray, Very Slippy Weather, etching, 25.2 × 19.4 cm, 1808.
Source: The Trustees of the British Museum, London.

of buyer and seller are undertaken but also the social roles of seeing and being seen become relevant.[11]

Furthermore, caricatures were displayed and viewed in taverns, brothels, and smoking- or pool-rooms. People who had the financial means to buy caricatures usually kept them in portfolios or displayed them in a purposely arranged room within their house. In these male-dominated contexts it was principally men who frequented such

Figure 10.3 Richard Newton, Holland's Caricature Exhibition, drawing, 44.7 × 67.7 cm, 1794.

Source: The Trustees of the British Museum, London.

rooms. Caricatures aroused interest as well as moral objections within English society. In more conservative families, women were prohibited from viewing caricatures. The following citation from John Corry, made in 1803, provides an impression of the moral doubts concerning caricatures: "[. . .] and it is an authenticated fact, that girls often go in parties to visit the windows of printshops, that they may amuse themselves with the view of prints which impart the most impure ideas."[12]

Caricatures were displayed in different places and viewed by people from different social ranks, even if there were some moral restrictions. However, even if those who passed a caricature shop could view the caricatures, it did not imply that they were able to understand the content. At this point it is worthwhile reiterating the salient points of a caricature: a caricature is a specific art form used to depict people as well as political or social events. It is characterized by stylistic elements such as deformation, exaggeration, and disassociation as well as simplification and the interplay of various associations. Furthermore, a caricature may be intended to amuse, insult, or criticize, and often refers to current events. In eighteenth-century England, caricatures were an integral part of a flourishing print industry. Caricatures profited from the relative freedom of expression that existed during the reign of King George III and the marked interest in politics demonstrated by large sections of the populace. To understand political caricatures one has to be cognizant of contemporary political events, one has to know the parties, the politicians, and other prominent figures. Accordingly, the disenfranchised members of society (notably women and the working classes) may ordinarily have had a weaker grasp of the details of caricatures, which required the viewer to detect nuanced allusions to political events.[13]

Returning to the subject of space, one final aspect is worthy of explicit mention: the production, distribution, and consumption of caricatures were not exclusively located in London. It is true that London was the trade center of caricatures, but the processes of the town were not detached from those of the countryside. Caricature shops also existed, for example, in Bath, Brighton, and Norwich. Usually caricatures on sale in these destinations were imported from London, but sometimes they were also produced in the counties. On occasion editors or shop owners traveled into the counties to find new and talented caricaturists and also to establish commercial relationships. Furthermore, caricatures spread throughout the country by means of networks of family relations or were transported by travelers or showmen.[14] To sum up, there was an active trade-off between the capital and the counties, demonstrating, yet again, the importance of the subject of spatiality in connection with caricatures, seen as a network of geographical, economic, and social aspects.

Imaginary Space in William Hogarth's Credulity, Superstition and Fanaticism. a Medley (1762)

The focus now shifts to the representation of space within caricatures, and as a result, the two-dimensionality of visual art becomes the center of attention. The multidimensional aspects of space are not left behind, however, because representation requires an object or a medium in order for the work to be viewed. Accordingly, there is a relationship between the viewer and the work of art. As Winfried Fluck argues, represented space needs the engagement of a viewer to generate meaning:

> Both literary and pictorial representations of space thus create not only a mental but an imaginary space; even where this representation may appear life-like, truthful or authentic, its actual status is that of an aesthetic object that invites, in effect necessitates, a transfer by the spectator in order to provide meaning and create an aesthetic experience.[15]

In the following analysis of the engraving *Credulity, Superstition and Fanaticism. A Medley* (figure 10.4) by William Hogarth, the focus lies on the following three questions: How is space created in the picture itself? What meanings are indicated by the arrangement of the figures? How is the viewer involved in the process of meaning making?

Figure 10.4 William Hogarth, *Credulity, Superstition and Fanaticism. A Medley, Etching and Engraving*, 37.9 × 32.8 cm (main plate)/52 × 32.7 cm (lettering plate), 1762.

Source: The Trustees of the British Museum, London.

Before commencing the analysis, a few words are required about the artist himself and his attitude toward caricatures: William Hogarth was a prominent and respected English artist of the eighteenth century.[16] Although he refused to be described as a caricaturist, he nonetheless had a marked influence on caricature production and graphic satire of his time, because he dissolved the sharp distinction between high and low culture and in his work one infiltrated constantly the other.[17] Hogarth claimed to differentiate between character and caricature, as he disapproved of the predominantly Italian understanding of caricature—one that deforms and exaggerates the physiognomic characteristics of a person. Hogarth involved himself closely with the social and political issues of his time, exposing social evils through his work by using techniques that are now regarded as characteristic of caricature.

In *Credulity, Superstition and Fanaticism. A Medley* Hogarth depicts the religious community of Methodists by characterizing it as credulous, superstitious, and fanatical.[18] One may ask, why did William Hogarth depict the Methodists in such a defamatory way? As Robert Paulson writes, there is little known about Hogarth's religious attitudes. In his work one may find traces of Puritanism and ideas of the Enlightenment.[19] Besides the critique of Methodism, he also criticized Catholicism and the Anglican Church. In *Credulity, Superstition and Fanaticism. A Medley* he particularly showed the aspect of superstition and the involvement of the Methodists in actual social events, as will be shown later in the article.

Coming back to the subject of space, it is significant that in this engraving space is categorized according to two aspects: above and below as well as inside and outside. This can be shown by the aid of two axes: one axis (see figure 10.5) begins at the figure of the preacher, descends to a clerk, before finally bisecting a women lying on the floor. This particular axis creates a spatial dimension within the engraving itself. It gives an impression of height: above and below. The three figures that dominate this axis are relevant for an understanding of the engraving.

The preacher in the pulpit wears a harlequin's costume under his gown, and beneath his periwig—that is flying away—a tonsure appears, that serves to imply a connection with the Jesuits. He holds in his hands two puppets: one a figure of the devil and the other a figure of a witch. The notion of superstition and fanaticism is manifested by this figure of the preacher. Additionally, the figure is exposed to ridicule by being adorned in a harlequin's costume.

Figure 10.5 William Hogarth's *Credulity, Superstition and Fanaticism. A Medley* with a diagonal axis.

The clerk in the middle of the picture, at the center of the axis, is supposed to be George Whitefield, who, along with the brothers John and Charles Wesley, was one of the founders of Methodism. It can be inferred that Hogarth's contemporaries would have been expected to recognize this likeness to Whitefield. The eighteenth-century viewer may also have recognized the woman lying on the floor who marks the end of the axis as Mary Toft, the "Rabbit Woman."[20] She represents the notion of superstition and credulity. In the year 1726 Mary Toft pretended to give birth to rabbits. Because of her story, doctors, journalists, and a lot of curious people were in a dither, until the swindle was discovered.

The axis creates spatiality not only on the horizontal level (above and below) but also on a representational and figurative level as the line descends hierarchically from the religious specialists to the ordinary members of the religious community. As the figures are recognizable for contemporaries the axis is highly significant for understanding Hogarth's engraving and for situating the scene in the religious community of the Methodists.

As noted above, Hogarth depicts an interior and an exterior and does so on different levels. In the background of the picture are two windows, through which can be seen clouds and trees, and significantly, behind one window a man is standing. This man differs markedly from the other figures in the interior: he wears a turban, has a beard, and is smoking a pipe, attributes that are characteristic—or rather stereotypical—of the time for the depiction of a Muslim. Furthermore, his attitude differs from the other characters, because he seems to be relaxed, but attentive. He observes from the outside what is going on inside.

The motif of a Muslim as an observer of society had become commonplace in both European literature and visual arts since the famous French novel *Lettre persanes* (1722) of Montesquieu.[21] Through the eyes of Muslims or other outsiders the social evils of one's own society may be criticized. Hogarth uses this motif in *Credulity, Superstition and Fanaticism. A Medley* and it is remarkable that in this engraving he does not depict a black man, as he does in other works of art.[22] It may be argued that the choice of a Muslim was deliberate, functioning as an "other" to the religious scene depicted, such that his religious affiliation was more relevant. As Hogarth wishes to criticize a religious community, the figure of a Muslim emphasizes this religious aspect. The confrontation of the enthusiasm of the Methodists and a clearly positively marked Muslim was also the subject in the novel *The History of the Adventures of Joseph Andrews, And His Friend Mr. Abraham Adams* (1742) of Hogarth's friend Henry Fielding.[23] Hogarth expressed this literary criticism of Methodism and the positive characterization of a Muslim in his engraving subsequent to the novel of Fielding. Accordingly, one can say that the choice of depicting a Muslim as a positive figure in contrast to religious groups or social events of one's own society is embedded in the historical and literary tradition of Hogarth's time.

It is notable that Hogarth's Muslim is standing outside, outside of the room, outside of the Methodist community and, as a foreigner, even outside British society. From the position of the figure of the Muslim a second axis can be imagined, that projects outward to the viewer, who might be standing in front of the picture. The spatial differentiation thus becomes a new dimension: the space of the depiction is related to the space of the viewer. The representation leaves behind two-dimensionality and communicates with a viewer. Or, as the art historian Wolfgang Kemp would say, the artwork communicates with the viewer.[24] With this imagined axis, it can be argued, Hogarth asks the contemporary viewer to keep his or her distance from the Methodists, as the Muslim observer does. Using this particular method to create space, Hogarth makes a connection between the inside of the picture and the outside of the viewer.

The example of William Hogarth's *Credulity, Superstition and Fanaticism* shows that space can also be analyzed as an aesthetic aspect of caricatures. Within pictures, space can be created with the help of different axes that can be located either inside the two-dimensionality of the picture or connect the inside of the picture with the outside of the viewer. Meaningful spaces arise on different levels, be it inside the picture, or in relationship with the outside.

Conclusion

In this article the subject of space is related to caricatures and especially to caricatures from eighteenth-century England. Thereby two lines were traced: first, theoretical reflections about space and caricature were linked with geographical, economic, and socio-spatial aspects. I localized the caricature shops of London geographically. I connected this localization, on the one hand, to the economic phenomenon of the print industry and, on the other hand, to the social practice of shopping. I also mentioned specific rooms where caricatures were kept and displayed. These rooms could be either public or private and separated according to gender. It is characteristic of eighteenth-century English caricature that it was located in specific places and became part of a trading market. The correlation between the caricature and the print industry and its technical possibilities played a crucial role in the history of the European caricature. Concerning spatial differentiation, it has to be noted that there was a close connection between the location of the caricature shop, the quality and content of the caricature, and the social rank of the customer.

Second, I focused on representations of space and analyzed William Hogarth's engraving *Credulity, Superstition and Fanaticism. A Medley.* It is interesting to see how Hogarth creates space by making one distinctive axis and by situating prominent contemporary figures along this axis. Although the engraving contains an array of different elements and figures, the diagonal axis that divides the space horizontally into an above and below offers the main point of entry into the picture and provides the first impression of its content. Hogarth very obviously criticizes the Methodists and strongly relates them to superstition by referring to some actual events of his time. The category of inside and outside also served to generate some remarkable insights. The figure of a Muslim as an outside observer corresponds very closely with the position of the viewer: the Muslim figure is the one who most directly communicates with the viewer as well as connects the represented space within the engraving with the multidimensional space of the viewer. The figure of the Muslim is somehow like a linchpin, where different levels of spaces come together and create various meanings of the picture.

The relationship between space and caricature involves various aspects. I chose to focus on the caricature in eighteenth-century England, its space of distribution and consumption and its space of representation. Concerning the former aspect, further research might insightfully ask how new media changed the meaning of space, and with respect to the latter, it would be fruitful to focus on caricatures that distort spatial dimensions more so than in the present test case.

As the journalist Christian Hüttner wrote in the year 1806 in the German journal *London und Paris*, the caricature shops of London were always crowded.[25] Today a visitor to London might go to Little Russell Street in the area of Holborn—in the center of the capital, unlike in the eighteenth century when it lay to the north of the city (remember figure 10.1)—and visit the first cartoon museum that opened in 2006. And here is an insider's tip: no crowds!

Notes

1. For literature concerning the Mohammed cartoons controversy, cf.: Ursula Baatz, Hans Belting, and Isolde Charim, eds., *Bilderstreit 2006: Pressefreiheit? Blasphemie? Globale Politik?* (Wien: Picus, 2007); François Bœspflug, *Caricaturer Dieu? Pouvoirs et dangers de l'image* (Paris: Bayard, 2006); Bernhard Debatin, ed., *Der Karikaturenstreit und die Pressefreiheit, Wert- und Normenkonflikte*

in der globalen Medienkultur (Berlin: Lit Verlag, 2007); Jeanne Favret-Saada, *Comment produire une crise mondiale avec douze petits dessins* (Paris: Les Prairies ordinaires, 2007); Jytte Klausen, *The Cartoons That Shook the World* (New Haven, CT: Yale University Press, 2009); Risto Kunelius and Elisabeth Eide, "The Mohammed Cartoons, Journalism, Free Speech and Globalization," in *Reading the Mohammed Cartoons Controversy, An International Analysis of Press Discourses on Free Speech and Political Spin*, Risto Kunelius et al., eds (Bochum/Freiburg: projekt verlag, 2007); Monkia Glavac, *Der Fremde in der europäischen Karikatur, Eine Studie über das Spannungsfeld zwischen Belustigung, Beleidigung und Kritik* (Göttingen: Vandenhoeck und Ruprecht, 2012).

2. On censorship in nineteenth-century France see Robert Justin Goldstein, *Political Censorship of the Arts and the Press in Nineteenth-Century Europe* (New York: St. Martin's Press, 1989); Justin Robert Goldstein, *Censorship of Political Caricature in Nineteenth-Century France* (Kent, OH: Kent State University Press, 1989).
3. See Jon Stobart, Andrew Hann, and Victoria Morgan, *Spaces of Consumption, Leisure and Shopping in the English Town, c. 1680–1830* (New York: Routledge, 2007), 2.
4. Winfried Fluck, "Imaginary Space; or, Space as Aesthetic Object," in *Space—Place—Environment*, Lother Hönnighausen, Julia Apitzsch, Wibke Reger, eds., (Tübingen: Stauffenburg Verlag, 2004), 27.
5. Cf. Wolfgang Kemp, "Kunstwissenschaft und Rezeptionsästhetik," in *Der Betrachter ist im Bild: Kunstwissenschaft und Rezeptionsästhetik*, Wolfgang Kemp, ed. (Berlin: Dietrich Reimer Verlag, 1992), 22.
6. See Herbert M. Atherton, *Political Prints in the Age of Hogarth, A Study of the Ideographic Representation of Politics* (Oxford: Clarendon Press, 1974); Diana Donald, *The Age of Caricature, Satirical Prints in the Reign of George III* (New Haven, CT: Yale University Press, 1996); Cindy McCreery, *The Satirical Gaze, Prints of Women in Late Eighteenth-Century England* (Oxford: Clarendon Press, 2004); Glavac, *Der Fremde in der Karikatur*, Chapter 3.
7. See Stobart, Hann, Morgan, *Spaces of Consumption*, 26.
8. The representation of space in caricatures will be discussed in the following section. The caricatures that are presented here serve to provide basic information for the analysis of the relationship between space and caricature.
9. Cf. Stobart, Hann, Morgan, *Spaces of Consumption*, 20.
10. Stobart, Hann, Morgan, *Spaces of Consumption*, 16.
11. Cf. Stobart, Hann, Morgan, *Spaces of Consumption*, 17.
12. John, A. Corry, *Satirical View of London, Comprehending a Sketch of the Manners of the Age* (London: J. Morton, 1803), 148.
13. On the definition of the term caricature see *Grove Art Online*, s.v. "Caricature," by Judith Wechsler, et al., accessed September 7, 2010, http://oxfordartonline.com/subscriber/article/grove/art/T014063;

Encyclopedia of Aesthetics, s.v. "Caricature," by Michele Hannoosh, accessed September 7, 2010, http://www.oxfordartonline.com/subscriber/article/opr/t234/e0102. For further research on caricatures in eighteenth century England see: Donald, *The Age of Caricature*.

14. Cf. Donald, *The Age of Caricature*, 20, 204; McCreery, *The Satirical Gaze*, 25.
15. Fluck, "Imaginary Space," 25.
16. On William Hogarth's biography see Frederick Antal, *Hogarth and His Place in European Art* (London: Routledge & Kegan Paul, 1962); Ronald Paulson, *Hogarth, The Modern Moral Subject, 1697–1732: Vol. 1* (Cambridge: Lutterworth, 1992); Ronald Paulson, *Hogarth, High Art and Low, 1732–1750: Vol. 2* (Cambridge: Lutterworth, 1992); Jennifer Uglow, *Hogarth: A Life and a World* (London: Faber and Faber, 1997).
17. Cf. Donald, *The Age of Caricature*, 1–2.
18. For more information on the engraving *Credulity, Superstition and Fanaticism. A Medley*, see Bernd W. Krysmanski, *Hogarth's Enthusiasm Delineated, Nachahmung als Kritik am Kennertum, Eine Werkanalyse, 2 Bd* (New York: Georg Olms Verlag, 1996); Ronald Paulson, *Hogarth's Graphic Works* (London: The Print Room, 1989), 177–78; Glavac, *Der Fremde in der Karikatur*, Chapter 3.
19. Cf. Paulson, Hogarth, *The Modern Moral Subject*, 10.
20. Paulson, *Hogarth's Graphic Works*, 178.
21. Cf. Karl-Heinz Kohl, "Die Magie des Okzidents: Europa in den Berichten arabischer und chinesischer Reisender," in *Der Orient im Okzident, Sichtweisen und Beeinflussungen*, Irene A. Diekmann, Thomas Gerber, Julius H. Schoeps, eds. (Potsdam: Verlag für Berlin-Brandenburg, 2003), 79.
22. Cf. David Dabydeen, *Hogarth's Blacks: Images of Blacks in Eighteenth Century English Art* (Mundelstrup: Dangaroo Press, 1985).
23. Cf. Henry Fielding, *The History of the Adventures of Joseph Andrews*, Adam Potkay, ed. (New York: Pearson/Longman, 2010).
24. Kemp, *Kunstwissenschaft und Rezeptionsästhetik*, 22.
25. J. C. Hüttner, "Die jetzigen Karikaturzeichner in London," in *London und Paris, 1798–1815, hrsg. von F. J. Bertuch, Bd. 18* (1806), Rudolstadt: 7–10, accessed April 9, 2010, http://www.ub.uni-bielefeld.de/diglib/aufkl/londonparis/index.htm.

Chapter 11

In Search of Orientation in Cinematic Spaces: The Journey Motif in Fiction Film from a Transdisciplinary Perspective of Film and Religious Studies

Marie-Therese Mäder

This contribution is situated at the intersection of film and religion. Film and cinema each have the capability to create a highly meaningful space where the vivid and varying responses of their audiences are displayed. Film reconstructs many different spaces that are comparable to the meaningful places of religious traditions,[1] takes an active part in the process of transmitting (religious) locations, transforms existing (religious) areas into another sphere by its narrations, and creates new spatial orders.[2] The interaction between film and religion may be viewed from the perspective of cultural studies in the sense that both film and religion are part of a shared cultural field in which each of the two sectors may influence the other.[3] Contact and overlap between these two sectors may occur by means of a range of different processes and can be demonstrated in a critical analysis, such as that undertaken in this chapter. Stuart Hall articulates these cultural processes in his circuit of cultures: "In other words, the question of meaning arises in relation to *all* the different moments or practices in our 'cultural circuit'—in the construction of *identity* and the marking of difference, in *production* and *consumption*, as well as in the *regulation* of social conduct."[4] Here I focus on representation in film production and consumption in accordance with the reception process. Meaning arises in these two moments of the circuit of cultures according to the rules and conventions of film's aesthetic narrations, the way religious symbols and narrations are represented, reconstructed, and transformed and the various ways in which the audience receives, perceives, and interprets these filmic narratives.

The present contribution is articulated in four steps. First, the fundamental difference between filmic and cinematic space is introduced. Second, taking the example of a journey, I suggest how meaningful cinematic space may contain a religious dimension. I will explain how transformational processes and the search for orientation characterize cinematic journeys and how they are linked with religion. Thus, in a third step, I discuss the filmic qualities of the journey by considering the properties of filmic space and in the fourth step its effect on the spectators. The central questions are as follows: How does film communicate and link spatial and personal orientation processes? How does the protagonist's spatial experience and its filmic realization engage with the spectator's perception?

Cinematic Space—Filmic Space

Even if the filmic space depicts a 100 percent fictitious world, the reception process in the cinema remains an experience in the real world: we, as an audience, can get angry, we can fear, we cry, and laugh. In the following analysis I would like to think about meaningful space in cinema in order to focus on the reception process, particularly to consider how an audience's invitation to experience a film is guided by the film's aesthetic realization. A consideration of how diegetic space is received as a world of its own may also provide some input into how human beings give meaning to space in general.

Two different kinds of spaces may be discerned in cinema. The semio-pragmatic approach to film by Odin underscores the double process of the film-text construction. Two spaces are involved in this process: one takes place in the space of the text's production—in this case, the film—and another is realized in the reception process.[5] According to this model it is possible to identify the *filmic space,* which is constructed by filmic devices and based on a more or less three-dimensional setting.[6] The setting itself, the location of the shooting in or outside of the studio, is of course a further important space relating to a film and it exists in the profilmic area, which may be added to Odin's model. The fictitious worlds of the *filmic space* are projected and presented on a two-dimensional screen. During the screening, in the reception process, the audience transforms the two-dimensional images back into a three-dimensional world. Consistent with Odin's model, this space of reception may be called the *cinematic space.* This space is dependent on the audience's film-reading skills, its previous knowledge, and imaginative power. So, when *filmic space* has to be perceived three-dimensionally and decoded by the audience, we can say that *cinematic space* only exists in the spectator's imagination. One important means of realizing three-dimensionality in film is by employing movement. The camera, the movement of the characters, and/or the staging of objects produce this effect, most often, all three aspects working together. The effect of continuity is given by the editing and supports the impression of an ongoing motion.

To scrutinize how the spectators are involved in constructing meaningful cinematic spaces I will focus on filmic journeys. Journeys are a central motif of many film genres and subgenres and are basically connected with movement.[7] In contemporary fiction film the journey motif is often linked with the character(s)' search for existential orientation and meaning. Their search for meaning has different dimensions: on one level the journey is a narrative pattern, on another level there are specific aesthetic strategies to express this search for meaning, and on a third level the search is communicated to the spectators. The construction of meaningful space is crucial on all three levels. It is a field where cultural markers such as landscapes, architecture, cities, and (religious) symbols, both on screen and in the minds of the spectators, define meaningful space.

From Cinematic Filmic Journeys to Religion

Since the beginning of film history, journeys have represented a prominent motif in many film plots.[8] But what defines a journey with a religious dimension? In some rare cases the films deal with pilgrimages situated in a religious tradition like in *Bab'Aziz*,[9] which tells the story of a journey to a meeting of dervishes, or the setting of *Le grand voyage*,[10] which is situated within the Muslim pilgrimage Hajj to Mecca, or in *Kingdom of Heaven*,[11] a historical drama set in the time of Christian pilgrimage to Jerusalem. Yet, there is another way in which filmic journeys are afforded a religious dimension. If the motif is used to represent a process of transformation, in which the characters are searching for orientation in a double sense, both geographically and personally, a religious dimension characterizes the filmic journey. In this case the journey connotes ritual functions by which the characters undergo an individual transformation. It is not a coincidence that characters often die on filmic journeys. In eighty films out of a total corpus of 135 films on which this research is based, death is a central topic in the sense that a character dies on the journey or tries to overcome a person's death.[12] The filmic journeys may be seen as a configuration with conspicuous connections to religion and may be discussed as an intersection where the two cultural fields, film and religion, exchange and mutually influence one another with patterns, values, and narrations.

Furthermore, a transformational process can be seen as a fundamental quality of religious systems. David Shulman and Guy D. Stroumsa discuss this quality in the context of religious systems: "All religious systems offer ritual programs, more or less elaborate, of varying types, whose primary goals are either to transform the participants from one state to another or to mark such *transformation*."[13] Fritz Stolz also sees transformations as linked with orientation processes in religious symbol systems. In this case, orientation may be seen as a functional aspect of religion: "The heuristic principle is that all transformations, which are set in motion by regulated and detailed processes, achieve a common bias. A religious message disposes of a purpose, it provides human beings open minded by nature with orientation."[14] Related to cinematic journeys, it thus follows that a transformational process in a filmic journey can provide orientation to the traveler. If this is the case, his or her search for orientation may have a religious dimension.

In both of these interpretations of the transformation process, a functional approach is used to define religion. Most authors in the field of film and religion choose this perspective on religion, as is the case with Conrad Ostwalt who emphasizes the sensory impact that films have on religion: "However, to understand religion functionally, we are concerned not with its truth claims but rather with how it orders life and functions within individual and cultural frameworks."[15] Ostwalt and many other authors in the field such as John C. Lyden,[16]Daria Pezzoli-Olgiati,[17] and Brent Plate[18] refer to the functional definition of religion devised by Clifford Geertz.[19]

There are genres and subgenres, in which a journey figures as either a prominent aspect or as the central motif, notably the ethnographic film, travel films, road movies, and many science fiction films. The journey motif is central in each of these named (sub-) genres, but always in a specific way, regardless of whether a transformational process is present. Similarities can be identified with respect to the aesthetic properties of the journey motif in such films. In the case of transformational processes, aesthetics plays a central role in the representation because its function is communicating the search for orientation. The audience is enabled to experience, to witness, or even to feel sympathetic toward the character who is being transformed.

This ability represents an essential condition of the reception process: the audience is required to follow the narration emotionally in order to understand journeys that depict a transformational process. In the next step I will emphasize some of the aesthetic qualities that relate to the journey as a process of transformation in filmic spaces and how they affect the audience. The aesthetic of the filmic representation constructs a religious dimension that is communicated by emotions and affects that engage both the body and senses of the audience.

Sensory Space

A character's search for orientation on a journey is often mirrored in the filmic setting: in the landscapes, cities, and buildings. Filmic parameters such as the camera, editing, and sound design shape and construct the filmic space, which guides the orientation process in a spatial fashion. The spectators take an active part in this process from a privileged position. In the following considered examples I analyze different aspects of the interplay between the audience and the film, focusing on the journey as part of an orientation process that turns the cinematic space into meaningful space.

There are two crucial aspects of the filmic journey that influence the motif in its narrative configuration and aesthetic appearance. The first aspect refers to the *relationship between space and time*, and the second aspect is concerned with the interplay between the sensual representation of the journey and the *involvement of the spectators*. The space and time dimension of the journey is constituted by the structure and the aesthetics of the narration. "Space and time are the subjects of the travelogue; indeed travel rhetoric frequently combines, or collapses, the two."[20] This quotation from Jeffrey Ruoff indicates the importance of these two dimensions. The focus in the narration often lies on the spatial and less on the temporal representation. The filmic narration favors space perhaps because of its ability to shape it visually by the photographic accuracy of the settings or by the construction of space by filmic devices such as the editing process or camera movement. The spectators reconstruct the spatial dimension of the narration, in the reception process, in two specific ways. According to Edward Branigan the audience immediately experiences the aesthetics of the surface—including sound design and dialogs—in a bottom-up process during the film's reception:

> Some perceptual processes operate upon data on the screen in a direct, "bottom-up" manner by examining the data in very brief periods of

> time (utilizing little or no associated memory) and organizing it automatically into such features as edge, color, depth, motion, aural pitch, and so on. Bottom-up perception is serial and "data-driven," and produces only short-range effects.[21]

At the same time, the spectators follow the structure of the filmic narration in a *top-down process*. "Other perceptual processes, however, are based on acquired knowledge and schemas, are not constrained by stimulus time, and work 'top-down' on the data, using spectator's expectations and goals as principles of organization."[22] The narration, for example, may contain gaps to create a density of events or to achieve suspense.[23] The spectator must utilize the acquired skill to follow the narrative structure and to construct a story from the plot presented.[24]

A crucial means of manipulating time and space is the editing process: sometimes, far-off places are connected by juxtaposing images from one take to the next. Filmic space may not even exist in the real world; it may be a unique digital construction, or be a set built by production designers. Further, orientation within filmic space also depends on how the different takes are joined.

So, film is sometimes characterized as a medium with metaphysical qualities with respect to representing time and space, because the diegetic world is not bound by any physical or "logical" laws. More than a hundred years ago, gestalt psychologist Hugo Münsterberg already pointed out the temporal possibilities of film. In the following quotation time can easily be replaced by space as is indicated by the brackets:

> The photoplay, on the other hand, does not and must not respect this temporal [spatial] structure of the physical universe. At any point the photoplay interrupts the series and brings us back to [spaces in] the past. [. . .] With the full freedom of our fancy, with the whole mobility of our association of ideas, pictures of the past flit through the scenes of the present. Time [Space] is left behind. Man becomes boy; today the [spatial] here is interwoven with the day before yesterday [place in the past]. The freedom of the mind has triumphed over the unalterable law of the outer world.[25]

The *physical universe* is independent of the filmic representation, so that the construction of cinematic space may be understood as a recreation of or a reflection on human space and time. Münsterberg's quotation does not suggest that filmic space can be edited with no restrictions at all. On the contrary, filmic space within narrative films has its own set of rules; these narrative films are produced in the Western hemisphere and in many cases are based on the Hollywood continuity system, which has been film-historically generated and has led to a highly conventionalized editing and staging system to represent the unity of space and time.[26] Filmic space and time are encoded by specific clues that communicate temporal, spatial, and causal relations. The audience has to decode these filmic codes to comprehend the intended relations—to understand a film is less of a "given" than it might appear to be; it is part of our acquired cultural knowledge.[27]

The bottom-up process presupposes an audience's ability to perceive what is immediately shown on screen. This aspect of the reception process is basic to understanding spatial properties and relations. It completes cognitive comprehension and film-reading skills with sensory perception as an additional aspect in the reception of filmic spaces.

I would like to highlight three types of spatial properties that have a particularly important impact on the audience's sensual experience during screenings. The first property is the texture of the surface, its three-dimensional quality, and its cultural peculiarities as captured in the vegetation, landscape, architecture, and cinematic construction of inner spaces, cities, and villages. Often, this texture of the surface is represented by photographic accuracy referring to the material character and spatial dimensions as they are known in the real world. The second property is concerned with motion produced by a mobile frame, and the third combines motion with emotion as an important means of involving the audience in the narration process.

The involvement of the audience in capturing the texture of the surface and exploring the cultural peculiarities of the imagery is crucial in the following scene from *Bab'Aziz-Le Prince Qui Contemplait Son Âme* (Nacer Khemir, CH/HU/FR/DE/IRN/TN/GB, 2007).[28] The female protagonist Ishtar (performed by Maryam Hamid), a young girl, and her grandfather Bab'Aziz (performed by Parviz Shahinkhou) discover a mysterious mosque after a long walk together through the desert. He is very tired and falls asleep soon after arrivalat the mosque. She, on the contrary, is curious and explores the mosque by perceiving the texture, its three-dimensional quality, and the cultural peculiarities of the building. Her entrance into this religious space through the curtain with calligraphic signs changes Ishtar's mood completely; overwhelmed by the sensual stimulation she undergoes a transformation.

The curtain separates two very different spaces, the one in front of the curtain belongs to the everyday world even if there are already men

Figure 11.1 Bab'aziz—Le Prince qui contemplait son âme (00:48:02).

at prayer, seated on the floor and listening to the music. The space behind this separation is the world Ishtar is going to explore, taking the audience with her. First we see how she perceives the impressive architecture, texture, and surface of the building (figure 11.1).

In the following shot the calligraphy of the mosque's ceiling is shown in a close-up that attracts and captures the girl's marvelling gaze. The letters are arranged in a curved spiral-like pattern causing a pull effect. It attracts the spectators' gaze providing an optical illusion whereby Ishtar's impression is paralleled.

The religious space is represented by music, its rhythm, the clear voice of the singers off-screen and the movement of the dervish. The audience explores the architecture of the building (figure 11.2) and the calligraphy on the wall through the girl's eyes.

The men are in the process of leaving the mosque and only Bab'Aziz remains, sleeping in a corner, together with a dancing dervish. Ishtar is attracted by the sound of a woman's chant offscreen and enters deeper into the mystical sphere of the musical source. Ishtar meets three women veiled in blue sitting in a dark cove of the mosque and opens the veil of the singer in the middle. The opening marks the entrance into a world of sensual and intimate experience. The movement of the veil is graphically arranged as a wipe, opening a new space of experience.

She listens to the singer and exchanges glances with her. We see the face of the singer in an extreme close up. In the reverse shot the extreme close up of Ishtar follows, her eyes are closed and there

Figure 11.2 Bab'aziz—Le Prince qui contemplait son âme (00:48:09).

is a smile on her lips, while she is listening and moving her head slightly to the music. The girl's ecstatic state signals an internalization of her sensuous space of perception. She is obviously delighted and entranced. The editing pattern goes back and forth seven times during the musical performance.

Afterwards Ishtar leaves the mosque and passes by a dancing dervish. She stops and observes his dance; then the camera changes to a point of view (POV) shot of the dervish. The POV is a specific means of expressing what a character experiences from his/her own subjective perspective.[29] A fast 360° degree camera movement shows the windows of the inner room and simulates the rotation experience of his dance. In the scene presented the audience is enabled to follow the dervish's rotation in order to perceive the rotating surface of the ceiling. The turning camera renders the ceiling in a blurred quality imitating the dervish's movement. The shot contrasts with the beginning of the sequence in which the calligraphy was presented in a sharp and close manner. Therefore the space of the current scene is perceived as a movement, whereas in the beginning it is shown as a static view.

The girl is now outside the mosque and watches the dance of the dervish from a roof window of the mosque. This shot helps to orientate the audience in filmic space to enable them to understand Ishtar's POV that follows. The light of the main room shines on her face and illuminates it. Again we see how Ishtar is fascinated and attracted to this place as she is almost peeping into the mosque. Her contagiously enchanted behavior lets this place become special and mystical.

At the end of her exploration the camera changes the perspective once again to a POV shot that shows what the curious eyes of Ishtar see. She is fascinated by the dynamic of the dervish's dance and the audience participates in this privileged observer position from a high-angle shot. The light of the window through which Ishtar is watching casts a beam of light on the dancing dervish. Its color has changed from a light yellow to a bright blue. Then the light is darker and announces the end of Ishtar's nightly adventure in the mosque. The reverse shot shows how Ishtar is gazing through the window and provides a final, last perspective of the mosque.

This very bodily and sensitive experience of the religious setting is communicated by visual and aural markers as well as by motion. The audience is invited to explore the surface of the mosque sensitively; the camera slowly pans across the walls and allows the spectator's glance to explore the texture of this meaningful space. The feeling of three-dimensionality is given by the movement through space; in this case the characters as well as the camera are in motion. In addition, the meta-diegetic music,[30] partly staged within the scene and partly a distinct musical score, which is only recognized by the audience and not by the figures, influences the atmosphere of the specific cultural setting. Ishtar is attracted by the sound as its volume increases and follows it to its source.

The connection between motions and emotions are important means to construct religious meaningful space. How they affect the audience is discussed in the following.

Motions and Emotions

The mobile frame involves the spectators by using motion. It may be produced by a mobile camera, a camera for example fixed onto a trolley with tracks or wheels—the dolly—or by fixing the camera onto a moving object like a car or a train. The movement of the camera can be realized in a tracking shot either by a pan, a crane shot, or a tilt, which produces frames that scan the space vertically. This movement of the camera three-dimensionally may occur forward, backward, laterally, upward, downward, or any potential combination of these spatial directions.

The motion produced by a mobile frame simulates motion for the spectators. The reception experience evokes a kinaesthetic reaction and may be called an *affective sensation*. The audience joins the moving images and gains a feeling of flowing through cinematic space. This kind of mobile perspective is called *phantom ride* and is a means of immersing the audience in the scenic events by stimulating their kinaesthetic senses, akin to the rollercoaster ride films shown in funfairs and amusement parks.[31] The audience perceives mobile frames in an immediate bottom-up process which allows them to anticipate the movement. The camera motion may indicate how much the viewers are invited to witness the journey affectively.

As the preceding remarks have shown, the journey motif may trigger kinaesthetic affects by a mobile subjective camera. In addition to affective effects, the movement of the journey may also be connected with emotions. The emotional style and effect of a journey represents another property, which influences the reception process. According to Noël Carrol while it may be thought that affects triggering emotion are dependent upon certain values, it is also possible to regard emotional affects as immediate kinaesthetic reactions:[32]

> Once we perceive the objects of our attention under the criterially relevant categories—such as harmfulness *and* disgust with regard to horror fictions or the improbability of the morally correct outcome with respect to suspense—the intended emotion is apt to be raised in us. That is, as a result of entertaining the appropriate appraisals—usually the ones implicated by the criterial prefocusing—we are likely to undergo the bodily responses, such as laughing, crying or squirming in our seat, that the movie makers have planned.[33]

There is of course no one unique emotional purpose of a journey.[34] The analytical challenge is to decode the emotions linked with the journey. Emotions can be conveyed by the protagonist's psychic shape, but also by the score, specific situations, and atmosphere.[35]

Giuliana Bruno discusses this relationship between motional affects and emotions in *Atlas of Emotion*. She goes so far as to argue that motion principally evokes emotion: "I will set forth as a major premise of the atlas, *motion, indeed, produces emotion* and that, correlatively, emotion contains movement."[36] Bruno elaborates the interchanges between architecture, geography, and film. She states that film mediates the movements of persons and things through the camera, and at the same time, film moves through an inner space of imagination, memory, and the topography of affects. Bruno further considers film and architecture as visual media; referring to this, she reflects upon architecture and film as haptic matters and tries to show how both fields trigger a tactile path in their spatial representation.[37] Her argument helps to show how meaningful spaces in cinema are constructed by staging motion connected with emotions. Both can be perceived during a screening and they represent a way of giving meaning to space in cinema that can be experienced bodily.

A central scene in *Mar Adentro* (Alejandro Amenábar, ES/FR/IT 2004) represents an appropriate example of a mobile frame that connects the journey motif with both properties of motion and emotion. A *travelling*, a technical term to describe a mobile frame, is used to express the protagonist's psychic shape, his wishes and fears. The film centers on the question of euthanasia, that is to say mercy killing, which is illegal in Spain, and is based on a true story: Ramón (performed by Javier Bardem) who spent seven years as a mechanic on a ship traveling the seas has been paralyzed for twenty-six years as a result of a swimming accident. He can only move his head, while he fights with legal procedures to die with dignity and to take his own life. It is not worth living, being confined to his bed, depending completely on other people. The impossibility of moving through space by cruising the sea, triggers his wish to die. A lawyer, Julia (performed by Belén Rueda), supports his cause. Both feel a tender but impossible love for each other. Julia is married and Ramón cannot bear that he is not able to fulfill this love bodily. Meanwhile Rosa, a local woman, strives to convince him that life is worth living. At the end she accepts his will to die and helps him to finish his life.

The scene I discuss deals with the journey on an imaginary level and has the function of visualizing for the audience the reason that underlies the protagonist's wish to die, expressed by means of a specific movement: Ramon imagines in a daydream that he can fly. He is dreaming that he leaves his bed and flies out of his window to the seaside. On a narrative and stylistic level the journey is employed as a means of symbolizing vitality and self-determination.

The scene starts with the crippled Ramón lying in his bed; suddenly he is able to move his hands. He stands up and we see his feet touching the floor. At the beginning he moves cautiously, one foot after the other. Getting more and more confident, Ramón pulls his bed a little further to the side in order to take a running start (figure 11.3) to jump and to fly out of his window. The reverse shot shows his subjective view before he leaves the window. The window functions as subframe in which on both sides hang white curtains covering one third of the frame. The open middle part gives a clear view of the landscape, and the curtain a blurred one through the fabric. The open window signals the start of his flight into open space. Almost without effort he seems to overcome gravity. Only his loud and intense breathing in the soundtrack reveals physical labor.

His flight to the beach is a traveling shot from a high-angle camera and is recorded in fast motion. The sequence is uncut, that is, filmed as a long take. The recording in fast motion marks the situation as a

Figure 11.3 Mar Adentro (00:40:32).

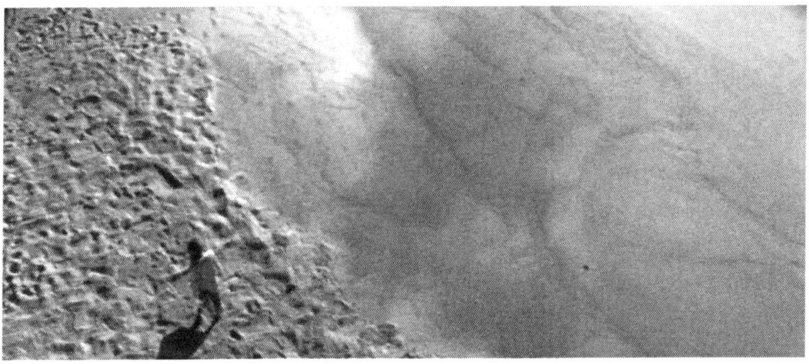

Figure 11.4 Mar Adentro (00:41:56).

dream or a vision. The traveling pan carries the audience along with Ramón out of his room to the sea. In his dream and thereby hinting at the impossibility in his "real" life he is able to rise and to fly out of the window. In his imagination, Ramón is free like a bird; he flies weightlessly over the landscape. The music supports this optimistic dream with the popular aria from the opera *Turandot* by Giacomo Puccini. The lyrics of "vincéro" mean, "I'm going to win," and correspond to the images of the sea. The music score indicates that Ramón is going to "win," which means in his case the fulfilment of his wish to die.

Before the sea appears, the camera pans up to the sky, and stops behind Julia. First Julia is seen from a high-angle perspective (figure 11.4) and then at eye level.

He meets Julia at the beach, touches her shoulder and says before he kisses her: "I've heard you're here. That's why I flew to you."

The editing starts again with the passionate kissing. Julia and Ramon are filmed in a close-up. It is somehow a return from the previous pure fantasy trip. Before his accident this situation would have been possible.

Because the audience joins the flight with Ramón from his room to the beach, we leave his home as well. The fast motion recording evokes a sensation of real flying. One could almost say that the spectators are leaving the cinema as soon as the screen opens up and takes them along over the landscape. *Mar Adentro* invites the audience to take part in Ramon's reality and in his dream experiences.

This example demonstrates how affects and emotions are closely intertwined with motion and how they are mutually dependent within the cinematic space. The movement over the landscape with its trees, hills, the beach, and the sea is perceived, on the one hand, as a dream-like situation and on the other as a very realistic scene. The sensation of flying can be experienced affectively and at the same time the audience feels sympathetic toward Ramón's psychic condition.

Meaningful Space in the Field of Film and Religion

An audience may experience religious filmic space spatially and emotionally, because the cognitive act of watching a film is accompanied by a sensual mode. The journey motif may represent a spatial search in such a way that the audience can participate physically in the protagonist's search for orientation and in the transformational process through emotions and affects. The top-down cognitive act of understanding the plot and transforming it into a story is accompanied by a sensual mode, which is communicated by a bottom-up process.

Affects and emotions are closely intertwined and in respect to the cinematic journey, they are mutually interdependent. The examples considered reflect the filmic strategies of constructing and interpreting religious space. The spectators are involved in experiencing the two-dimensional space and decode it as a three-dimensional world in their imagination. This means that they are deeply involved in this process on a cognitive, affective, and emotional level and have to be aware of their physical responses to merge into this process.

The three-dimensionality of the filmic journey is provided with emotional and affective qualities, which enrich space in cinema sensuously and let the audience experience space as meaningful. But to give meaning to space always implies an active participation by the spectators who are looking for values in the situation represented;

according to Odin it thus follows that "the subject joins the adventure and expedition which implies the search for values."[38] For a successful trip the recipients need sensibility and imaginative power to apprehend the narration and to construct meaningful spaces.

The examples considered represent only incomplete points from which to scrutinize the complex correlation between meaningful space, film, and religion. But it shows that the immediate perception in the bottom-up process is central to the reconstruction of meaningful space. Meaning always has something to do with human communication as audiences bring their background, cultural knowledge, and values into the cinema to join the process of spatial meaning making. On the other hand films bring emotionally and physically meaningful spaces into the spectator's imagination. They enable the spectators to see and to explore spaces they may never otherwise see in real life. The cinema in this case provides a public space for religiously connoted experiences and the representation of religious phenomena.

Notes

1. Melanie J. Wright, *Religion and Film: An Introduction* (New York: I. B. Taurus, 2007).
2. Charles Martig and Daria Pezzoli-Olgiati, eds., *Outer Space: Reisen in Gegenwelten* (Film und Theologie 12; Marburg: Schüren, 2009).
3. Marie-Therese Mäder, "A Cultural Studies Approach to Film and Religion, Context and Film Analysis of YES (Potter, GB/USA 2004)," 101–18 in *Approaches to the Visual in Religion,*Christopher Rowland and Daria Pezzoli-Olgiati, eds. (Göttingen: Vandenhoeck & Ruprecht, 2011).
4. Stuart Hall, *Representation. Cultural Representations and Signifying Practices* (London: Sage Publications, 1997), 4.
5. Roger Odin, "Kunst und Ästhetik bei Film und Fernsehen," in *Montage A/V, Zeitschrift für Theorie und Geschichte audiovisueller Kommunikation 11/2/2002* (Marburg: Stoke, 2002), 42.
6. Important exceptions are, for example, the animation film, in some cases the experimental film and computer animated scenes.
7. My research is based on a corpus of 135 fiction films, most of which were produced in the last ten years. The majority of the films can be categorized as art-house films, and a smaller group belong to the American independent cinema. The rest of the films are (co-)produced in Europe, North Africa, and the Middle East. See Marie-Therese Mäder, *Die Reise als Suche nach Orientierung, Eine Annäherung and das Verhältnis zwischen Religion und Film* (Marburg: Schüren, 2012).

8. Annette Deeken, *Reisefilme: Ästhetik und Geschichte* (Filmstudien 38; Remscheid: Gardez! Verlag, 2004); Winfried Pauleit, Christine Rüffert, Karl-Heinz Schmid, Alfred Tews, and Bremer Symposium zum Film, eds., *Traveling Shots, Film als Kaleidoskop von Reiseerfahrungen* (Berlin: Bertz und Fischer), 2007.
9. *Bab'aziz. Le Prince qui contemplait son âme* (Nacer Khemir, CH/HU/FR/DE/IRN/TN/GB, 2007).
10. Ismaël Ferroukhi, MA/FR, 2004.
11. Ridley Scott, USA/GB/ES/MA/DE, 2005.
12. Mäder, *Die Reise als Suche nach Orientierung*, Chapters 1 and 3.
13. David Shulman and Guy S. Stroumsa, "Introduction: Person, Passages, and Shifting Cultural Space," in *Self and Self-Transformation in the History of Religions,* David Shulman and Guy S. Stroumsa, eds. (Oxford: Oxford University Press, 2002), 6. Emphasis original.
14. Fritz Stolz, "Effekt und Kommunikation," in *Religion und Rekonstruktion,* Daria Pezzoli-Olgiati, ed. (Göttingen: Vandenhoeck und Ruprecht, 2004), 114. The original German text is: Als heuristisches Prinzip gilt, dass alle Transformationen, welche durch die im Detail regulierten Vorgänge in Gang gesetzt werden, eine Gesamttendenz haben. Eine religiöse Botschaft insgesamt hat ein "Ziel" sie gibt dem von Natur aus offenen Menschen eine Orientierung [translation by the author].
15. Conrad Ostwalt, "Teaching Religion and Film. A Fourth Approach," in *Teaching Religion and Film,* Gregory. J. Watkins, ed. (Oxford: Oxford University Press, 2008), 37.
16. John C. Lyden, *Film as Religion: Myths, Moral, and Rituals* (New York: New York University Press, 2003), 44–48.
17. Daria Pezzoli-Olgiati, "Film und Religion. Blick auf Kommunikationssysteme und ihre vielfältigen Wechselwirkungen," in *Religious Turns—Turning Religions. Veränderte kulturelle Diskurse—Neue religiöse Wissensformen,* Andreas Nehring and Joachim Valentin, eds. (ReligionsKulturen 1; Stuttgart: Kohlhammer, 2008), 54.
18. Brent S. Plate, ed., *Representing Religion in Worldcinema: Filmmaking, Mythmaking, Culture Making* (New York: Palgrave Macmillan, 2003), 5.
19. Clifford Geertz, "Religion as a Cultural System," in *The Interpretation of Cultures: Selected Essays* (New York: Basic Books, 1973), 87–125 : (1) "A system of symbols which acts to (2) establish powerful, pervasive, and long-lasting moods and motivations in men by (3) formulating conceptions of general order of existence and (4) clothing these conceptions with such an aura of factuality that (5) the moods and motivations seem uniquely realistic (90)."
20. Jeffrey Ruoff, ed., *Virtual Voyages, Cinema and Travel* (Durham, NC: Duke University Press, 2006), 19.
21. Edward Branigan, *Narrative Comprehension and Film* (New York: Routledge, 1992), 37.

22. Branigan, *Narrative Comprehension and Film,* 37.
23. Branigan, *Narrative Comprehension and Film,* 37.
24. The Russian Formalists of the 1920s coined the terms *fabula* and *syuzhet* in film theory. David Bordwell uses *story* and *plot* for the same distinction: "Plot[:] In a narrative film, all the events that are directly presented to us, including their causal relations, chronological order, duration, frequency, and spatial locations. Opposed to *story,* which is the viewer's imaginary construction of all events in the narrative." David Bordwell and Kristin Thompson, *Film Art: An Introduction,* 8th ed. (New York: McGraw-Hill, 2008), 480; David Bordwell, *Narrative in Fiction Film* (Madison: University of Wisconsin Press, 1985), 48–63.
25. Hugo Münsterberg, *Das Lichtspiel: Eine psychologische Studie (1916) und andere Schriften zum Kino,* Jörg Schweinitz, ed. and trans. (Wien: Synema, 1996).
26. Bordwell and Thompson, *Film Art*; Kristin Thompson, *The Classical Hollywood Cinema: Film Style and Mode of Production to 1960* (New York: Columbia University Press, 1985), 194–213; Knut Hickethier, "Zur Analyse des Visuellen, des Auditiven und des Narrativen," in *Film- und Fernsehanalyse* (Stuttgart: J. B. Metzler, 1996).
27. Cf. Branigan, *Narrative Comprehension and Film,* 1992: "In order to connect our understanding of film to our understanding of the ordinary world, we have to be sensitive to the specific techniques available in film for representing space, time, and causal relation" (48).
28. *Bab'Aziz—Le Prince qui contemplait son âme* tells the story of the blind dervish named Bab'Aziz and his granddaughter Ishtar. They wander the desert in search of a great reunion of dervishes that takes place just once every 30 years. He does not know where it will take place and they walk without geographic orientation through the desert. He is convinced that God will guide him and his granddaughter to the right place. Bab'Aziz relates an ancient tale to keep Ishtar entertained. The journey is declared as a search for God. The question remains how this intended interpretation is transmitted to the audience. The search for orientation is represented by a loss of orientation on two levels: the main characters are wandering through the desert, simply guided by their faith, that God will show them their path. Also the audience cannot follow or cognitively understand the plot's changes in time and space because the film purposely lacks such markers.
29. Murray Smith, "Imagining from the Inside," in *Film Theory and Philosophy,*Richard Allen and Murray Smith, eds. (New York: Oxford University Press, 1997), 412–30.
30. Bordwell and Thompson, *Film Art,* "Diegetic sound: Any voice, musical passage, or sound effect presented as originating from a source within the film's world" (478). "Nondiegetic sound: Sound such as mood music or narrator's commentary represented as coming from space outside the narrative" (480).

31. Tom Gunning, "Traveling Shots," in *Traveling Shots, Film als Kaleidoskop von Reiseerfahrungen,* Winfried Pauleit et al., eds. and Bremer Symposium zum Film (Berlin: Bertz und Fischer, 2007), 24.
32. Noël Carrol, *The Philosophy of Motion Pictures* (Malden: Blackwell, 2008), 147–55.
33. Carrol, *The Philosophy of Motion Pictures*, 160.
34. Linda Williams, in her paper "Film bodies: Gender, Genre, and Excess" (Film Quarterly 44.4 [1991]: 2–13), discusses this part of my argument. She works out typical body responses toward different film genres such as horror films, pornographic films, and melodramas.
35. Carrol, *The Philosophy of Motion Pictures,* 147–60.
36. Giuliana Bruno, *Atlas of Emotion. Journeys in Art, Architecture, and Film* (New York: Verso, 2002), 6.
37. Bruno, *Atlas of Emotion*, 6.
38. Roger Odin, "Kunst und Ästhetik," 45. (Translated by the author). The original German text is: "auf eine Forschungsreise und ein Abenteuer ein, die aus einer Suche nach Werten besteht."

Chapter 12

Clothing as a Meaningful Marker of Space: A Comparative Approach to Embodied Religion from a Cultural Studies Perspective

Anna-Katharina Höpflinger

Clothing can be connected with various meanings and functions: we wear clothes to protect, construct, and form our body, as well as to communicate.[1] This communication proceeds in the sense of a "viscourse"—a discourse through visual codes.[2] A vestimentary viscourse forms part of sociocultural actions and complex identification processes. Nowadays, such a communication-theoretical approach to clothing is broadly accepted in apparel research, as Fred Davis already affirmed back in 1985: "That the clothes we wear make a statement is itself a statement that [...] has virtually become a cliché."[3] Following Malcolm Barnard, I define clothing as a complex communication system that embraces the whole of human appearance.[4] Clothing is interlinked with fundamental mechanisms of orientation in social networks. Human apparel is coupled with processes that allow agents to navigate their way through social interactions and include, for example, the rapport of gender, differing categories of age and social status, as well as varying ideas of beauty.[5] Besides the social construction of meaning on the level of a viscourse, clothing invariably contains a spatial element as well.[6] The human figure moves through space in clothing and thereby interacts with it.[7] But the category of space also plays a significant role in constructing the body itself,[8] as well as social and religious differences and hierarchies indicated through the spatial aspects of clothing, as I will endeavor to show in this chapter.

Clothing, Space, and Religion

Clothing may literally interact with space, in the sense of physical extension, as the following caricature by the French artist Charles Vernier illustrates (figure 12.1).

The caricature with the title "La crinolonomanie" depicts a noble British lady (perhaps even Queen Victoria) in her crinoline dress.[9] She is waiting to enter the ballroom while two men with picks try to enlarge the doorway. The sentence under the picture says: "AU CHATEAU DE WINDSOR: Comme quoi il est beaucoup plus facile d'élargir les portes que de diminuer les jupes" (At Windsor Castle: it is much easier to enlarge the doorway than to reduce the dresses).

Vernier mocks the contemporary fashion of his time by drawing attention to the spatial problems caused by the crinoline.[10] Thereby the French caricature not only creates boundaries distinguishing British and French fashion, but also hints at the social element of spatiality. The main figures belong not only to different genders, but

Figure 12.1 Charles Vernier, Caricature "La crinolonomanie," Le Charivari (1857), 25.7 × 37.3 cm.

Source: Staatliche Museen zu Berlin—Kunstbibliothek.

also to different social classes. The dress can be taken as a fashion critique as well as a criticism of the imperial attitude of richer (and royal) British people. The image captures the spatial and social relation between the figures depicted, characterizing the obvious imbalance at both the spatial and social levels.[11]

By looking a little more deeply into such interactions between clothing and space, it becomes useful to distinguish between different aspects of interdependencies: space can be important for the physical representation of clothing, as the caricature by Vernier shows, and it can also be seen as a topographical factor defining the clothing, as in the example of hiking boots, which help us to traverse rougher ground.[12]

Clothing can be a marker of social space and can construct and transmit hierarchies,[13] as indicated again in the crinoline caricature shown above; one has to be able to afford a garment that consumes so much fabric and room. A further example of social interconnections between clothing and social space is the ballroom (the intended goal of the lady in the crinoline). A ball is a spatial and fashionable *mise-en-scène* with sharp social demarcation processes and a special type of costume created for this occasion only. The lines between demarcation and belonging constructed by clothing are in this case social and spatial at the same time.

Taking a broader look at the boundary-making processes established through clothing, however, may also be seen to provide a sense of regional and social identity. Clothing may indicate geographical borders and areas of belonging, just as it does in Vernier's caricature,[14] where the French caricaturist mocks British royal fashion (even though the crinoline fashion originated in France). Another example of a vestimentary drawing of national-spatial boundaries is provided by traditional costumes. Such folk-costumes not only indicate a geographical area, but also represent an idealized imagery of the inhabitants of that region.

Nonetheless, clothing is not only a means of socio-geographical identity and idealization, but also a means of spatial and social marginalization, as we can see, for example, in an image of a tattooed seaman in indigenous garments.[15] The engraving, made by an unknown artist, portrays the seaman Joseph Kabris who was shipwrecked in the Marquesas Islands, and who, over time, adopted the apparel and appearance of the indigenous inhabitants.[16] The engraving stylizes him both as a marginalized European and as a noble savage. The little textile garment he is wearing and the replicated tattoos identify the man as belonging to a specific geographical and social marginal region, while at the same time bestowing upon him an exotic air. The theoretical categories of center and periphery may prove helpful in endeavoring to situate this "central" yet "marginalized" figure.[17] Additionally the idealized landscape depicted as background in this engraving should not be ignored, as it supports the dichotomy (re) presented by the clothing.

On the basis of these very different examples it is apparent that interactions between clothing and space can be manifold and that when analyzing the meaning of clothing it is advantageous not to ignore the category of space. From the great variety of possible interdependencies between clothing and space I have singled out three which are especially relevant for the present article: a geographical-topographical aspect, a hierarchic social-spatial aspect, and a spatial identity producing aspect.[18]

To link these thoughts to religion and to define the "meaningful" aspect of the interrelations between clothing, religion, and space, I follow Fritz Stolz's definition of religion, because in his work spatial aspects are significant categories used to analyze religious phenomena.[19] Stolz sees religion as a comprehensive, sense-generating system of orientation that problematizes the tension between controllable and uncontrollable aspects of human life:

> Everywhere, mankind is confronted with the task to order and control a world which is open and defies determination; we encounter everywhere powers beyond our control [......]; this is the place where religious problems emanate. The point is to give form to the uncontrollable, thereby making it approachable. That way, something uncontrollable is transferred into a controlled area, while its character stays intact; religion thus performs a depiction of the uncontrollable life dominating powers, while at the same time accepting the uncontrollable way of life emanating from it. Thereby offering mankind an essential and comprehensive system of orientation.[20]

As was the case for the interactions between space and clothing, so also interrelations between religion and space can be manifold, as the chapters in this book show.[21] Utilizing the three categories mentioned above, the following concrete examples will further demonstrate the point. On a geographical-topographical level, for example, mythologies can explain uncontrollable characteristics of landscape and create thereby religious meaning. The monster Typhon, slain by the God Zeus/Jupiter and captured beneath Mount Etna, is for example a religious interpretation of the volcanic activity of that Italian mountain.[22] But religious hierarchical systems are also often expressed by means of spatial aspects. In antiquity, spatial access to sacred sites, such as temples, was carefully restricted and hierarchically ordered,[23] while in some European Christian communities, religious specialists continue to be separated from the rest of the community during rituals, for example by preaching from a pulpit. On a spatial identity producing level we can stress for example that religion was, at least in some historical eras, constructed as tied to a specific geographical context and therefore intertwined with collective identification processes.[24] The sixteenth-century motto *cuius regio, eius religio* (whose realm, his religion) is an example of such constructions. Historical and contemporary transnationalization processes show that a clear relation between geographical and religious settings could never be taken for granted; they uncover complex boundary-creating processes and identification models.[25]

As these selected examples show, interrelations between clothing and space as well as between religion and space are manifold. In the following part I illustrate this connection between clothing, religion, and space by an example from Switzerland. Thereby I focus only on two of the above-mentioned three heuristic categories: the hierarchic socio-religious-spatial aspect, and the spatial identity producing aspect.

Religious and Regional Belonging: The Scaletta Cloak

Historical Context and Methodological Approach

In the Swiss canton of Grisons the so-called Scaletta cloak is used. It is a special cassock for Protestant pastors, and represents a peculiarity among Protestant clerical robes in Europe.[26] The garment is mostly cut as a semi-circle or three-quarter circle and fitted with a collar. It is worn open at the front over a black suit (for men) or Sunday best (for women) and is fastened with two pieces of cord directed under the armpits and tied at the back. Until the middle of the twentieth century it seems that the cloak was usually closed at the front and not at the back (figure 12.2). The Scaletta cloak is (and always was) black and nowadays—because of practical reasons—mostly made of a wool-synthetic mixed fabric. The traditional material was wool.

The Scaletta cloak developed from being an item of everyday man's outdoor clothing during the late seventeenth and early eighteenth century to a garment for male office bearers and can be found in this usage in different regions of Switzerland. A very similar cloak is depicted, for example, in the pattern book of Salomon Erb, a master

Figure 12.2 The pastor Ch. Gillardon during a festival celebrating 300 years of liberation of the Prättigau (a region in Grisons) from Austria, 1949 in Küblis.

Source: Ch. Höpflinger-Gillardon.

tailor from Berne, written and painted in the years 1723 or 1724.[27] The cloak depicted in this tailor's book is meant for the bailiff and is cut in a similar way to the contemporary Scaletta cloak, with the fabric used also noted as wool. In the nineteenth century, this sort of clothing—the focus is here exclusively restricted to the black version—was related to the official vestment for office bearers at funeral services.[28] Until the end of the nineteenth century, Protestant (male) pastors from Grisons wore the cloak with white bands at the collar, whereas the coffin bearers wore a cloak without bands. The combination of a black cloak and white bands can already be found in portraits of pastors from Grisons dating from the eighteenth century.[29]

Nowadays, in Grisons, the Scaletta cloak is worn by both male and female Protestant pastors when preaching sermons in the above specified way,[30] unless the wearing is unregulated, as a female pastor points out: "[. . .] For us it is optional [to wear liturgical garments]. That is why one can see all kinds."[31]

The data for my research have been taken from an empirical study focused on the Scaletta cloak that began in 2009. Twelve qualitative-empirical interviews with different agents (pastors, pastor aspirants, relatives of pastors, and community members) have been conducted and analyzed on the basis of Paul DuGay's five categories of the so-called *circuit of culture*: production, consumption, representation, identity, and regulation.[32] Additionally, I observed several services as a participant and, furthermore, could gain—with special reflection on all the methodological problems such an intervention brings with it[33]—some insight into the production of this special type of cloak. In the progress of my academic work I have been asked to sew two Scaletta cloaks, and at least one of these requests came because the pastor wished to have a tailor with a special historical interest in the cloak.[34]

The name of the cloak already indicates a relationship to a particular spatial context and links the cloak specifically to funeral services: the name "Scaletta" is probably taken from the Scaletta burial site in Chur, the capital city of Grisons.[35] But the naming of the cloak is not its only connection to spatial aspects. Two other striking interrelations are evidenced by a closer analysis of this garment.

First, the cloak can be seen as a visual and "meaningful" representation of a specific religious space connected, at least from an etic study-of-religions perspective, with socioreligious hierarchies and separations. Typically, the religious space represented by the Scaletta cloak is also clearly defined at a physical level. In this type of spatial setting, the Scaletta cloak identifies the religious specialist and thus allows for privileged socioreligious interaction.

Figure 12.2 shows a pastor from the region of Prättigau in 1949 during his speech at a festival to celebrate the tercentenary of the liberation of this region from Austria. He wears the Scaletta cloak to represent his role as a religious specialist as well as to demonstrate his connection to this specific geographical region. So secondly, as figure 12.2 demonstrates, the Scaletta cloak is tightly connected with the Canton of Grisons. At least since the late nineteenth century, the cloak is a sign of geographical belonging, and, from an emic perspective, can be expressed as a "tradition of Grisons."[36] Thus, on this level, we can distinguish processes of interaction between clothing, geographical identification, and idealization of a geographical area.

The Cloak Representing and Constructing Religious Space

As mentioned above the Scaletta cloak represents and at the same time helps to construct a specific sort of religious space qualified through a strongly performative aspect. This space is often (but not always) also clearly defined physically: normally the rituals marked by the Scaletta cloak take place in buildings with religious connotations. As I have observed, the pastors usually dress themselves with the cloak in the church building directly before the preaching service, and they also undress again in the church after the service. But often this dressing ritual depends also on the spatial possibilities of the specific church. M1, a male pastor, explains this as follows and links the dressing in the church with the change of (religious) role, whereby he stressed the importance of religious reflection during the process of changing:

> The moment to dress and the moment to undress; I don't choose them very purposefully. The problem is I'm missing a room to dress, something like a vestry, where I can go and where the cloak is hanging, where one can change dress [. . .]. I'm coming from here [= the parsonage], and I have thought about dressing here and going like that to the church [. . .], that would probably be more practical than to dress somewhere in the church. [. . .] But I try to dress in the church because I thereby reflect on what I'm doing, [. . .], I concentrate. And then I tell myself: now I'm dressing myself with the cloak, [. . .] now my task is needed, now the ministry is needed. I dress myself consciously.[37]

Sometimes the cloak doesn't belong to the pastor, but to the parish and is kept in the church, but this has some disadvantages, as the interviewee M12, a male pastor, makes clear: "I didn't want to borrow one that doesn't fit."[38]

To outline this connection between clothing and representing religious space, I've asked the interviewed pastors if there are special events or moments when they explicitly choose not to wear the Scaletta cloak. The answers were very different; most linked the wearing of the cloak to the sort of church service that they were practicing. But one interviewee, again the male pastor M1, made a clear spatial connection:

> Exceptions, when I don't wear the cloak? I do not wear it on the Alps.[39] [. . .] It is simply not practical enough. For the preaching service it would make no difference, but it would mean that I needed two sets of clothing and that would entail having to change and that is impractical. After taking off the cloak you are not yet suitably dressed for the Alps [in the suit you are wearing].[40]

This quotation shows that the construction of religious space does not depend on wearing this special sort of cassock, and that practical aspects—in this case regarding topographical difficulties—can be more important than a vestimentary marking of religious space or the religious specialist. The latter is nevertheless an important factor in religious clothing:[41] hand in hand with the representation of religious space goes a labeling of the religious specialist, as again M12 points out: "The cloak has [. . .] the function of a marker. When 500 people meet at a funeral service and half of them are dressed in dark colors, the cloak makes this clear: this is the pastor, he guides (not: commands)."[42]

The marking of the religious specialist is connected with "protection" and professionalism, as M1 explains:

> The pulpit presents a kind of protection, and perhaps one could say the same about the cloak. It offers protection to both sides, as it also protects the people. [. . .] They know that the things spoken by the pastor while he is in the pulpit are spoken in the pulpit. Thus, you see, that way, whatever is said is less direct, [. . . not so] face-to-face, therefore, I can position myself at different degrees of distance or closeness regarding what I am saying.[43]

These quotations make clear that the Scaletta cloak can (but does not have to) stand for the construction and representation of a certain religious "meaningful" space and a specific religious specialist. Thus, it provides orientation in a religious space of action and denotes a certain religious role as well as indicating a religious hierarchy.

The Cloak as a Marker for Geographical Affiliation

Besides a connotation with the construction of religious space, the cloak also represents, in an ideological way, the Protestant tradition of a specific geographical region (figure 12.2). The male pastor M12 declares:

> The cloak stands for the pastorship in Grisons and its identity, that is to say its distinctiveness. And it is a pity that some of my colleagues here don't like the cloak. To my knowledge the cloak is only worn in Grisons. The east Hungarian Protestants have some sort of similar cloak, but less plain. Therefore this cloak is part of a lived regionality, one of the greatest advantages of Switzerland.[44]

As M12 points out, the cloak is connected with the imagined drawing of geographical and religious boundaries. These processes of belonging, and on the other side demarcation, are interpreted very differently from an emic perspective. While M1 (August 16, 2009), like M12 above, stresses that the cloak is "a nice tradition of Grisons," M5 (September 28, 2009), a former pastor of Grisons is of the opinion that we find here a sort of "local patriotism in a negative sense."

Because nearly 40 percent of the pastors working in Grisons are originally from Germany, it is especially interesting to look on their strategies to cope with this local tradition.

A female pastor originally from Germany who wants to take a position in Grisons answers in this way: "I had a service here in Grisons and I wore my [Prussian] preaching robe. I was as a pastor from Germany on that occasion. But when I will work here, I have to decide if I will wear my preaching robe—my old identity—or if I adopt the local tradition."[45]

M12, also German, points out that questions regarding ethnic identity are multilayered: "I am German. Before I came to Switzerland I was rather critical towards Germany [. . .], but now I have discovered my German identity. But I love Switzerland and the rich tradition here [. . .]. And Grisons is again a little bit different [from the rest of Switzerland] and has a special, long history of its own. It's my home now, and I am someone from Grisons now. A hybrid sort of identity."[46]

This vestimentary connection to Grisons is not only linked to the question of how one is to dress when becoming a pastor in this canton, but it is also striking that pastors who leave Grisons no longer wear the Scaletta cloak in their new context, even if they have always worn it in Grisons.[47]

So, the Scaletta cloak is an interesting example showing that clothing can be used to construct and represent spatial mechanisms of belonging, and also, at the same time, of demarcation.[48]

Conclusion

In the case of the Scaletta cloak, on the one hand clothing represents a religious sphere of action and is a "meaningful" sign for a specific religious role in this religious space. On the other hand, clothing can be a marker of a geographical affiliation within a specific religious tradition. Regarding the first aspect, clothing is more flexible with regard to its tangible position, while on the second level it is bound to a collectively specified (and imagined) geographical area. Both aspects of the clothing-space relation have to do with processes of identification, which go hand in hand with belonging and demarcation mechanisms.

The example of the Scaletta cloak shows how useful it is to link historical and social aspects of religious clothing to spatial ones.[49] Regarding this special sort of Protestant cassock, space is at the same

Figure 12.3 Anonymous, Bündner Kalender, 1860.

Source: Bündner Kalender, Casanova Druck und Verlag.

time constructed as historical and social: the cloak, cut in a historical manner, marks social hierarchic interactions in a clearly defined religious space and reminds one at the same time of historically emergent aspects of geographical belonging. As this example shows, all these fields of meaning are "sounding" through a viscourse based on a strong material aspect. As we have seen, clothing is first and foremost a material medium of communication. Therefore the meaningful representation and construction of space occur not only through the context and action of "wearing the garment," but are based on the textile, its fabric, pattern, cut, and flow.[50] Clothing itself is a spatial entity.

I began my argument with a vestimentary caricature of Vernier to show this physical-spatial aspect of clothing. I would like to end with a very similar caricature connecting this physical-spatial aspect explicitly to religion and morals by mocking vanity and fashion in a religious sense (figure 12.3). But in contrast to Vernier's ballroom, the consequences of wearing overly spatial garments are different. While the men with the picks in Vernier's caricature can enlarge the doorway, such a solution is not possible in the anonymous caricature printed in the *Bündner Kalender*, a magazine from Grisons, from 1860 (figure 12.3).

We can see in figure 12.3 some ladies in crinoline dresses seeking entry into heaven. The caricature is entitled "The crinoline at heaven's gate." The ladies are addressing Saint Peter who is guarding the door: "Oh, Mister Peter, wouldn't you kindly make way for us?" But Peter answers: "I'm sorry, my ladies, but the path into heaven is too narrow for you."

Notes

1. Regarding clothing and communication see, for example, Gregory Stone, "Appearance and the Self," in *Dress and Identity,* Mary E. Roach-Higgins, Joanne B. Eicher, and Kim K. P. Johnson, eds. (New York: Fairchild Publications, 1995), 19–39; Alison Lurie, *The Language of Clothes* (London: Bloomsbury, 1992); Ruth P. Rubinstein, *Dress Codes: Meanings and Messages in American Culture* (Boulder, CO: Westview Press, 1995); Jennifer Craik, *The Face of Fashion: Cultural Studies in Fashion* (New York: Routledge, 1994); Stefanie Würtz and Roland Eckert, "Aspekte modischer Kommunikation," in *Inszenierungsgesellschaft. Ein einführendes Handbuch,* Herbert Willems and Martin Jurga, eds. (Wiesbaden: Westdeutscher, 1998), 177–91.
2. The term "viscourse" goes back to Karin Knorr Cetina. She used the term to describe the interplay between visual depictions and their embedding in a continuous discourse focusing on the question

"how do visual codes structure physical sciences?" See Karin Knorr Cetina, "'Viskurse' der Physik. Wie visuelle Darstellungen ein Wissenschaftsgebiet ordnen," in *Konstruktionen Sichtbarkeiten,* Jörg Huber and Martin Heller, eds. (New York: Springer, 1999), 245–63. The term is understood in the present article as a communication at a visual level, connected with normative aspects and identification processes.

3. Fred Davis, "Clothing and Fashion as Communication," in *The Psychology of Fashion,* Michael R. Solomon, ed. (Lexington, MA: Lexington Books, 1985), 15–27 (15).
4. Malcolm Barnard, *Fashion as Communication* (New York: Routledge, 2008).
5. Anna-Katharina Höpflinger, "'Mehr verschandelt als verwandelt': Kleidung als Medium der Geschlechterkonstruktion in religiösen Symbolsystemen," in *Handbuch Gender und Religion,* Anna-K. Höpflinger, Ann Jeffers, and Daria Pezzoli-Olgiati, eds. (Göttingen: Vandenhoeck und Ruprecht, 2008), 243–55.
6. Anna-Brigitte Schlittler and Katharina Tietze, *Kleider in Räumen* (Winterthur: Alataverlag, 2009).
7. See Gabriella Schubert, *Kleidung als Zeichen: Kopfbedeckungen im Donau-Balkan-Raum* (Wiesbaden: Harrassowitz, 1993).
8. See Thomas A. Tweed, "Space," *Material Religion* 7.1 (2011): 116–23 (120–23.).
9. Crinoline, made of horsehair, was invented in the 1830s by a Frenchman, Monsieur Oudinot-Lutel. In the 1840s and 1850s a great range of different dress petticoats and all sorts of new crinolines were patented, the most famous of which was the steel cage crinoline (for patents regarding attire from 1846–1920, see Mihaela Cornelia Peteu and Sally Helvenston Gray, "Clothing Invention: Improving the Functionality of Women's Skirts, 1846–1920," *Clothing and Textiles Research Journal* 27 [2009]: 45–61). Crinoline was replaced in the late 1860s by the bustle dress. Regarding the crinoline fashion, see Anne Buck, *Victorian Costume and Costume Accessories* (Bedford: Carlton 1984), 27–40; Adelheid Rasche and Gundula Wolter, *Ridikül: Mode in der Karikatur 1600 bis 1900* (Berlin: SMB/ Du Mont, 2003), 231–32; Erika Thiel, *Geschichte des Kostüms: Die europäische Mode von den Anfängen bis zur Gegenwart* (Berlin: Henschel, 2000), 322–30.
10. See Rasche and Wolter, *Ridikül,* 231–43.
11. For interactions between clothing, gender, and space see Peter McNeil and Giorgio Riello, "The Art and Science of Walking: Mobility, Gender and Footwear in the Long Eighteenth Century," *Fashion Theory 9* (2005): 175–204.
12. For interrelations between shoes, space, gender, and social differences, see Peter McNeil and Giorgio Riello, *Shoes: A History from Sandals to Sneakers* (New York: Berg, 2006).

13. See Barnard, *Fashion as Communication*, 61–64, 104–16.
14. Schubert, *Kleidung als Zeichen*, 11.
15. The image is available online in the Alexander Turnbull Library, Wellington, New Zealand, No PUBL-0184–06, accessed December 26, 2013, http://mp.natlib.govt.nz/detail/?id=28143&l=en.
16. See Joanna White, "Marks of Transgression: The Tattooing of Europeans in the Pacific Islands," in *Tattoo. Bodies, Art and Exchange in the Pacific and the West,* Nicholas Thomas, Anna Cole, and Bronwen Douglas, eds. (London: Reaktion Books, 2005), 72–89.
17. See Fritz Stolz, *Grundzüge der Religionswissenschaft* (Göttingen: Vandenhoeck and Ruprecht, 2001).
18. See Thomas Schindl, *Räume des Medialen: Zum spatial turn in Kulturwissenschaft und Medientheorie* (Boizenburg: Werner Hälsbusch 2007).
19. See, for example, Fritz Stolz, *Weltbilder der Religionen. Kultur und Natur. Diesseits und Jenseits. Kontrollierbares und Unkontrollierbares* (Zürich: Pano, 2001); Fritz Stolz, "Ausbreitungsstrategien und Universalisierungstendenzen in der europäischen Religionsgeschichte," in *Europa verstehen,* Ingolf Dalfert, Hans J. Luibl, and Hans Weder, eds. (Zürich: TVZ, 1997), 39–63; Daria Pezzoli-Olgiati and Fritz Stolz, *Cartografia religiosa—Religiöse Kartographie—Cartographie religieus*e (Bern: Lang, 2000).
20. Stolz, *Grundzüge*, 33. In the German original: "Überall steht der Mensch vor der Aufgabe, seine Welt, die offen und nicht festgelegt ist, zu ordnen und zu kontrollieren; überall ist er mit Mächten konfrontiert, die sich dieser Kontrolle entziehen [. . .]; an dieser Stelle sind die religiösen Probleme angesiedelt. Es geht darum, dem Bereich des Unkontrollierbaren eine Form zu geben, mit der sich umgehen lässt. Dabei wird einerseits Unkontrollierbares in die Kontrolle übergeführt, andererseits aber doch wieder belassen; Religion leistet also eine gleichzeitige Darstellung der unkontrollierbaren lebensbestimmenden Mächte und der unkontrollierbaren Lebensordnung, die darin gründet. Dadurch ergibt sich eine grundlegende und umfassende Orientierung des Menschen."
21. See, for example, Tweed, *Space*; Pezzoli-Olgiati and Stolz, *Cartografia religiosa;* Danièle Hervieu-Léger, "Space and Religion: New Approaches to Religious Spatiality in Modernity," *International Journal of Urban and Regional Research* 26.1 (2002): 99–105; Roger W. Stump, *The Geography of Religion. Faith, Place, and Space* (Lanham, MD: Rowman and Littlefield, 2008).
22. See Anna-Katharina Höpflinger, *Schlangenkampf: Ein Vergleich von ausgewählten Bild- und Textquellen aus dem griechisch-römischen und dem altorientalischen Kulturraum* (Zürich: TVZ, 2010), 158–63.

23. See Anna-Katharina Höpflinger, "From Visibility to Invisibility: A Semiotic Approach to Marduk's Battle against Tiamat on a Relief from Palmyra," in *Approaches to the Visual in Religion*, Daria Pezzoli-Olgiati and Christopher Rowland, eds. (Göttingen: Vandenhoeck and Ruprecht, 2011), 39–50.
24. See Stomp, *Geography of Religion*.
25. See Stolz, *Ausbreitungsstrategien und Universalisierungstendenzen*; Janine Dahinden, Anna-Katharina Höpflinger, and Anne Lavanchy, eds., "Gender and Religion: Towards Diversity," *Special Issue of Women's Studies: An Interdisciplinary Journal* 41.6 (2012).
26. Ralph Kunz and Thomas Schlag, "Des Pfarrers neue Kleider... Erkundungen und Erwägungen zur Talarfrage in den reformierten Landeskirchen," *Pastoraltheologie* 98.6 (2009): 212–32 (217).
27. Quirinus Reichen and Karen Christie, eds., *Das Schnittmusterbuch von Salomon Erb, "Livre des Chefs d'Oevre de la Maistrise des Tailleurs de Berne, 1730"* (Bern: Bernisches Museum, 2000), 16–17.
28. Hans Berger, *Evangelisch Chur—Seine Prädikanten, Kirchen und Friedhöfe* (Chur: Calven, 1978), 157–58.
29. See J. Jürgen Seidel, *Die Anfänge des Pietismus in Graubünden* (Zürich: Chronos, 2001), 101, Figure 5.
30. Bündner Synodale, *Der Scalettamantel*, 17.
31. Interview with W2, October 25, 2009, translated by AKH.
32. Paul DuGay et al., *Doing Cultural Studies: The Story of the Sony Walkman* (London: Sage 1997), 3–4.
33. Special thanks to Annegret Kestler and Jacqueline Grigo for discussing these problems with me.
34. E-Mail from W6, female pastor in Grisons, December 11, 2009.
35. Berger, *Evangelisch Chur*, 157–58.
36. Interview with M1, male pastor in Grisons, August 16, 2009.
37. Interview with M1, August 16, 2009, translated by AKH.
38. Interview with M12, December 10, 2011, translated by AKH.
39. In summer sometimes regular Sunday services are held on the Alps.
40. Interview with M1, August 16, 2009, translated by Daniela Casanova.
41. See Linda Arthur, *Religion, Dress and the Body* (Oxford: Berg, 1999); Linda Arthur, *Undressing Religion: Commitment and Conversion from a Cross-Cultural Perspective* (Oxford: Berg, 2000).
42. Interview with M12, December 10, 2011, translated by AKH.
43. Interview with M1, August 16, 2009, translated by Daniela Casanova.
44. Interview with M12, December 10, 2011, translated by AKH.
45. Interview with W5, October 25, 2009, translated by AKH.
46. Interview with M12, December 10, 2011, translated by AKH.

47. Interviews with W3, May 22, 2010.
48. See Arthur, *Undressing Religion*, 4.
49. Edward W. Soja, *Thirdspace: Journeys to Los Angeles and Other Real and Imagined Places* (Cambridge: Blackwell, 1996), 53.
50. See Lou Taylor, *The Study of Dress History* (New York: Manchester University Press, 2002), 3–23, 24–63, 72–74.

Chapter 13

In Quest of the Known: Multiple Spaces as an Essential Component of Traveling Experiences from a Social Sciences' Perspective

Tommi Mendel

> *"Yeah. Those two were on the pilgrim's route. They had guidebook written all over them. If not, like I already said, we'll deal with it when it happens." He pulled a knot of hair out of his beard. "You know, Richard, one of these days I'm going to find one of those Lonely Planet writers and I'm going to ask him, what's so fucking lonely about the Khao San Road?"*
>
> Alex Garland, *The Beach*[1]

It has never been as easy to travel the world as it is today and there were never as many travelers hitting the road for their own sake as there are nowadays. Whereas in previous eras only wealthy people could afford to go abroad to see the proclaimed tourist highlights, in the last few decades a vast number of people have been leaving home for several weeks or even months on end in order to undertake extensive travel to all parts of the planet.

Relevant pull factors for this fairly new leisure activity are the uncomplicated accessibility of most traveling destinations due to open borders, simple immigration procedures, and inexpensive airfares combined with low-budget living costs, extensive and comfortable infrastructure of most destinations, the range of transport options available, and the general safety of the places visited. Such incentives and opportunities for increased travel have been accompanied by the enormous rise of travel agencies and outdoor-activities shops, fueled by advertisements by the travel industry, as well as manifold media exposure: travelogues of famous people as well as social network sites and traveling blogs readily accessible on the Web. Push factors have also had a concerted influence, notably, increasing individualization in Western societies, financial affordability and material affluence of the participants, the development of leisure time, and the social approbation as well as the feasibility of taking "time-out" from work.[2] The steadily growing number of independent travelers has created commonly used, beaten tracks in various world regions—the "pilgrim's routes" mentioned in the quote at the beginning of the chapter—designed by the popular guidebooks. The same is true for more overt pilgrimages.

The findings of my own extensive field research undertaken along the pilgrimage routes around *Shikoku* in Japan and *Santiago de Compostela* in Spain, as well as along traveler's paths in Cambodia, Laos, and Thailand show striking similarities between foot-pilgrims and backpackers from postindustrial societies with regard to the historical development of the two modes of travel, the participants, their preconditions and motives, as well as the structure, practice, and form of being on the road.[3] The studies also clearly evinced that the participants were influenced and driven by collective images created before their departure and that these images endured throughout the journey.

In this chapter I shall try to demonstrate how combinations of multiple spaces are responsible for the perceived images and experiences of a contemporary traveler and how geographic spaces interfere with the mental spaces and vice versa in the course of a journey.[4] This chapter is based on the data derived from the above-mentioned field studies, mainly gained through observant participation and narrative interviews, as well as from the findings of scholars in the field of pilgrimage and tourism studies.

First, I will define the term "traveler," which will be used throughout, before outlining the structure of a journey consisting of ten steps. Thereafter, I will demonstrate how the perceived experiences of travelers are influenced and constructed by media inputs, with the help of empirical examples, before elaborating on a theoretical framework of the different geographic and mental spaces that a traveler may be confronted with.

Defining the Term Traveler

> How can all the other pilgrims—mostly not penniless people—deal with these horrific flophouses, in which every so often one becomes

> ill-treated [. . .] Nonetheless, well off people play the pauper in such contexts and 50 year old, chubby American ladies sentence themselves to walks of 40 kilometers or more in the scorching heat. How guilty must they feel to take all this on? (Hape Kerkeling: *Ich bin dann mal weg*)[5]

For the following discussion it is essential to first define the term "traveler," as there are limitless ways and forms of being on the road and the traveling phenomenon itself is in a constant state of flux due to historical, sociocultural, economic, and political factors. One can differentiate short excursions and vacations from longer journeys of several weeks as well as business or research trips from tourist or leisure activities, which equally need to be distinguished from working migrations or refugee movements. Hence significant determining factors are the motives and intentions of the protagonists, the traveling destinations, the time and modes of being on the way, the available budget and form of organization, the frequency and style of movement as well as the daily activities *en route*.

In this chapter I will use common denominators at the *etic* level to define the term "traveler," which are: leisure activities along a self-organized and flexible itinerary, frequent movement between destinations, a relatively low budget to spend, and a minimum duration of four weeks continuously spent on the road.[6] The latter two factors correlate because the low-cost travel destination allows even less well-off people to travel for more than a standard holiday length of a couple of weeks. The minimum time period of four weeks enables the protagonists to develop a self-identification as a traveler and to be recognized as such by their peers. Moreover, a minimal time away from everyday life is necessary to undergo a personal transformation, which is one of the most common motivations to undertake a journey of several weeks.[7]

The definition above includes pilgrimages. As I have shown elsewhere, pilgrimages are and always have been undertaken in combination with tourist interests, and tourist-motivated journeys mostly comprise religious places like sacred buildings or sites.[8] Additionally, contemporary scholarly discussions regarding pilgrimages include voyages to nonreligiously associated places such as the graveyards of famous people or war victims, disaster sites, plus monuments of cultural and national importance, which may also be on a tourist itinerary.[9] Pilgrimages, therefore, cannot be treated as separate from ordinary travel and are here understood as sub-categories of modern tourism.

Accordingly, in the context of the present paper, the definition of the term traveler will also include pilgrims.

It has to be pointed out, nonetheless, that today's travelers do not form an homogeneous group—the individual motives, interests, and activities on the road differ too much for such a monolithic construct. Additionally, my empirical studies show clearly the difficulties with which protagonists are commonly confronted in order to define themselves as travelers, at least within the first few days of their journeys.[10] On the other hand, it is striking that most of the protagonists departed with a firm idea of what a pilgrim or a backpacker is, informed by the role models of the medieval pilgrim or the early explorers, respectively. In this context, images of what a "real" traveler ought to be exist and persist throughout a present-day journey and help the traveler not only to separate him/herself from organized forms of tourism but also from the world back home. Before explicating these images, I shall briefly elucidate the structure of a journey, going beyond the spatial and temporal distinction between being at home and being *en route*.

A Journey's Structure

> Behind us lay the whole continent of America and everything Neal and I had previously known about life, and life on the road. We had finally found the magic land at the end of the road and we never dreamed the extent of the magic either. (Jack Kerouac: *On the Road*)[11]

Most theoretical and empirical studies relating to traveling focus solely on the time that is spent on the road, while only very few also include the aftermath of the journey.[12] To study the experience of travel properly, however, I think it is absolutely imperative to look at the entire traveling process, starting before the actual time on the road and going beyond the return back home. Thus the structure of a journey is crucial, not least because one has to bear in mind that the outer journey of a traveler always correlates with his/her inner journey; there is an inseparable link between the physical and the psychological journey.[13]

Following the approach of Victor Turner's structural concept of pilgrimages as rites of passage—the separation from the daily routine, the temporary and locally different reality on the road, and the return to every day life[14]—I propose to extend Turner's three-step structure to a ten-step model that also includes the time before and after the actual journey.[15] A journey involves more than simply the period

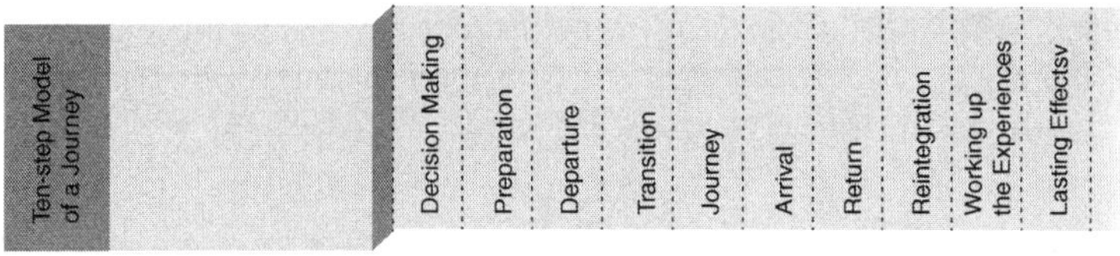

Figure 13.1 Graph, Ten-Step Model of a Journey.

of what Turner calls "the liminoid state";[16] it starts with the decision to hit the road and ends with the lasting effects after being back home for some time. In doing so protagonists undergo the journey on a geographical as well as a mental level, both are significant for the traveler's personal transformation process, which has neither a fixed starting point nor a fixed end point, in contrast to traditional rites of passage. The ten-step model for all types of traveling consists of the steps described in figure 13.1.

At each step on the journey the traveler is confronted with various mental and geographical spaces. In doing so the unconscious combination of these spaces is responsible for the traveler's perceived experiences of the journey.[17] Thereby the mental images, and as a consequence also the actually perceived experiences, are highly influenced and created by a number of media inputs as well as by the interests of the tourist industry, as will be outlined in the following section.

Media Inputs and Traveling Experiences

> And therefore I understand the passion of travelogues, their craziness and fraud. They provide us with the illusion of something not existing anymore but which ought to exist [. . .]. (Claude Lévi-Strauss: *Tristes Tropiques*)[18]

Media, such as literature, films, and the fine arts are connected to traveling in a twofold way: on the one hand they provide the possibility of breaking away from everyday routine and setting the stage for an escape into another world, and on the other hand they supply the stock of collective images and fantasies.

The German sociologist Christoph Hennig states that already in the early days of tourism travelers had been influenced by collective images derived from paintings and books when choosing places to visit. As a result, sites unmentioned in authoritative books or not depicted in the fine arts were not considered interesting and therefore not on the tourist itinerary.[19] Accordingly, what was important for the traveler was not the insight into the foreign countries visited, but rather the desire for "sensuous experiences of fictive spaces": well-known images from literature, art, and collective imaginations had to be found again in the material reality.[20] Consequently, tourists have always traveled and still continue to travel more to affirm these collective images and stereotyped depictions than to look for something new, as the comment of one of my informants, a Canadian traveler in his mid-twenties, suggests after having entered a traditional Hmong village in North-Laos: "That's why I came to Southeast Asia, to see people living like that."

Paul Gauguin's painting *There Lies the Temple* is an exemplary illustration of how collective images are created and how the desire to visit these "fictive spaces" may be triggered and amplified. It has to be said that Gauguin's initial image of Tahiti was of an exotic paradise far away from the influence of European civilization, and it was not until he arrived there in 1891 that he was first confronted with a reality that

Figure 13.2 Paul Gauguin: Parahi Te Marae Aka (There Lies the Temple), Oil on canvas, 1892, 26 × 32 inches.

Source: By courtesy of Philadelphia Museum of Art: Gift of Mr. and Mrs. Rodolphe Meyer de Schauensee, 1980, Inv. Nr. 1980–001–001.

failed to match up to his expectations. Nevertheless, in order to live up to the expectations of the collective images at home and to meet the expectations of the art-market in Paris, he continued to paint and write about events that either no longer existed, or in fact had never existed at all. In *There Lies the Temple* Gauguin implies that he encountered a monumental stone statue next to a temple. Historical research shows, however, that at the time Gauguin lived in Tahiti such temples and sacred figures were no longer in existence, and that in reality he copied the narratives of the nineteenth-century travel-writer Jacques-Antoine Moerenhout about the moon goddess *Hina* of the Easter islands.[21] This case study does not concern an artistic interpretation of a real occurrence but, according to art historian Christoph Becker, a deliberate falsification, "a game with an inadequate knowledge of the recipients" at the time.[22]

The image of the Noble Savage as an allegory for the losses resulting from civilization and the craving for an alternative lifestyle in a paradise with exotic palm beaches has entered into the collective imagination, desires, and fantasies of Western societies since the eighteenth century and continues to this day.[23] Contemporary media, from newspaper reports to TV documentaries, from travel-writers to tourist industry publicity brochures, websites and guidebooks, all deploy these collective images of the exotic paradise producing the desired results. In doing so, they deliberately distort reality in order to profit from the inadequate knowledge of the average recipient. The following quote from Lonely Planet's *Southeast Asia on a Shoestring* illustrates this clearly:

> This sleepy country of gilded temples, tangerine-robed monks, emerald rice paddies and impossibly friendly people is unlike any other [...] responsible, low-impact trips into its many protected areas allow you authentic adventures and sometimes very close encounters with the wild things that dwell in them! [...] its heart is still very much a shoestringer's gem, where rugged travel off the package circuit reaps the greatest rewards [...] thatched houses, colourful ethnic tribes and unbroken forests is much the same as that which greeted French colonials a century ago [...] Now really is the best time to visit, before mystical Laos inevitably slips into the 21st century.[24]

Narratives of travel-writers, web-blogs, and particularly detailed and comprehensive guidebooks influence travelers widely in their choice of travel destinations as well as in their traveling style, self-representations, and aspirations.[25] "With only a few lines, Lonely

Planet can make a place boom or let if fall into disgrace," writes Peter Welk referring to one of the most popular and widely sold guidebooks worldwide.[26] The following extracts of interviews with a resident and a guesthouse owner along the pilgrimage to *Santiago de Compostela* as well as a guesthouse employee in Laos elucidate the potency of famous people's travelogues and prevalent guidebooks:

> It's a fashion. It all began, well, Paulo Coelho made it very famous with his book about the Camino. A Brazilian. And it continued more and more. But in my youth, some 20 years ago, there wasn't one single pilgrim passing here, not one. Of course nowadays there are many passing by.[27]
>
> For example in the year 1999 there were many, many Brazilians passing by because Coelho wrote this book [...] For the past four years it's been the same with Germans. A very famous German did the Camino and also wrote a book about it. That's why there are a lot of Germans now.[28]
>
> In Vang Vieng there were only [...] three guesthouses when I was young, in about 1988. And now it has changed a lot in Vang Vieng. There are more than 200 guesthouses around Vang Vieng. So it's very busy, it's good for business, we get some money.[29]

The potency of guidebooks and travel-writing results not only in a general interest in certain places that travelers may choose to visit besides influencing the formation of prevalent routes within an established tourist infrastructure but it also creates a common view of how to travel the "authentic way." Although all of the travelers I came across during my field studies traveled in a relatively comfortable and safe way, the pilgrims of the medieval period and the early discoverers served as obvious role models. The "authentic" is clearly associated with images of the past, the exotic, the untouched, and the adventurous, being kept alive purposefully by the media and the tourist industry.

Studying log-records in various guesthouses and having talked to numerous travelers, I detected an identical set of vocabulary to that used in the popular guidebooks: "a must see," "authentic," "real experience," "well worth a visit," "incredible," "amazing," "unbelievable," "in the middle of nowhere," "to share the local's life," and so on. The same vocabulary influences the content of postcards and short travel reports by email or social networking services like Facebook, with the accounts mostly palliating the actual life on the road by having the collective images and the expectations of the receiver at home in mind. The media and information scholar Jutta Bertram states that travelogues resemble advertising texts and holiday photos replicate postcards or pictures of catalogues, in both cases it is an affirmation of known images.[30]

Photographic themes like the exotic (markets, food, simple housing, rice paddies etc.), the beauty of nature (sunrises, sunsets, waterfalls, mountains, beaches, etc.), the tourist highlights (buildings, natural sights, etc.), the adventure (arduous and extraordinary activities, etc.), the people (children, old and poor people, monks, etc.) as well as the proof of having been there (oneself—to show to friends and family at home) are popular among a traveler's hunt for already known subjects. I would like to enhance this point with the following interview excerpt of an informant looking at photos from other travelers a few weeks after the accomplishment of her journey to *Santigo de Compostela*, "It's nice to have a look at these pictures. Most of these pictures I had also taken myself and showed them to others already. These are the prototype pictures or the well known themes [. . .] It's hilarious, partly these photos are exactly the same as mine."[31]

As a result, travelers don't see the facts as such but rather the already existing images of the facts, based on their imaginations and projections as well as their collective desires and fantasies.[32] Since humanity is not able to envisage a nonspatial world, Hennig argues, these mental images always need to be localized in a space, he calls it "the geography of the imaginary."[33] As a consequence travelers are likely to witness common perceptions and experiences at certain stages of their journeys due to the various media inputs. The following interview excerpts from two German travelers in their mid-twenties recorded at two different stages on their pilgrimage to *Santiago de Compostela* exemplifies this thought:

> I think it was Astrid saying that the pilgrimage is divided into three parts: the first part is for the body, the second part for the spirit and the third part for the soul. I'm still within the first part, considering that, my body is doing very well.[34]
>
> Here along the Camino one lives through highs and lows. In the Meseta you walk 200 kilometers through the steppe, that's where you're confronted with your soul. Before, there is the part for your body until Burgos. In this part I realized how the body and the soul are connected: when I was in pain for several days the soul was suffering as well. But when you overcome everything, like here in Galicia, it's great. People say that the part for the body starts at the beginning of the pilgrimage either at St. Jean Pie-de-Port or in my case at the

> Somport pass. It's when the head is with your legs, when you feel and overcome your body. And then, when you leave Burgos and walk through the Meseta for 200 kilometers, it's the part for the soul. In the Meseta the path is straight on and flat, it's a wasteland. At the beginning it's beautiful but after a few kilometers it turns to the part for your soul, because many personal issues come to the surface. And here in Galicia, apparently the part of the spirit is about to begin. I'm excited about it.[35]

Independently of each other these two informants mention the division of the pilgrimage into three experiential parts, whereas the latter mentions the hardship of the *Meseta*, the stretch of the deserted plateau, and the concomitant emotional immersions. Interestingly enough quite a number of travelers I talked to in Spain mentioned these experiential divisions and came to know a comparable "spiritual experience" at the *Meseta*. Looking closer at one of the popular pilgrimage guidebooks, I detected the following self-explanatory line: "Meseta pure! The following stretch is one of the hardest and for many the spiritually most thrilling along the way."[36] Lines like this not only form collective images but also common experiences. For that reason in empirical research one should be alert to detect information and statements based on media inputs that become copied and passed on by hearsay, rather than taking the above declared "spiritual experiences" for granted like some scholars have done.[37] A long time in the field and closeness to one's informants help to sift out the real sensed experiences from the proclaimed experiences.

To summarize this section, the effective experiences and impressions during a journey commingle with collective images and personal associations brought along.[38] Or in the words of Hennig: "We travel in the real and the imaginary world at the same time when traveling as tourists *en route*."[39] Therefore I think it is well worth looking more closely at each of these different worlds, these "fictive spaces," and their composition.

Differentiation of Spaces

> After only a few hours in England, all the unpleasant parts of my trip were tumbling from my memory. Rationally, I could still just about weigh things up and remember that for the majority of the time I'd been miserable, but I felt so happy that I'd done it, and had survived, that my positive emotions were already beginning to swamp

> everything else. In my mind, the trip was turning itself into an amorphous good thing [. . .] All my contradictory feelings were passing through a filter which was picking out anything unpleasant or painful [. . .] My journey round India was already reducing itself into just another person's "amazing experience." (William Sutcliffe: *Are you Experienced?*)[40]

In order to distinguish and outline the different layers of space to be confronted with while traveling, I consider it useful to come back to the previously discussed general structure of a journey consisting of ten steps and take it as a base for the following theoretical model. As mentioned before, at each step of the journey a traveler faces different spatial layers whereby the combination of these layers is constitutive for the traveler's sensed experience.

As a distinction on a primary level I suggest differentiating between geographical or physical spaces and mental spaces. On a secondary level the geographical spaces can be subclassified into *Everyday Space*, *Tourist Space*, and *Local Space* whereas the mental spaces may be divided into *Imagined Space* and *Experienced Space*. I will begin by describing the three geographical spaces:

Everyday Space

The everyday space comprises the physical surroundings in daily life, like one's home, one's town, and one's place of work and leisure embedded in a familiar society and culture. Travelers live in the *Everyday Space* before their departure and again after their return from the journey.

Tourist Space

The majority of the travelers move along identical and pre-designed routes visiting places of interest for tourists either on foot or by means of transport provided for tourists. They eat and sleep in places specially created for tourists and as a result more of their encounters on the road are with other travelers than with local people. Therefore this space firmly enclosed within a tourist infrastructure must be differentiated from the *Local Space*. Travelers are confronted with the *Tourist Space* when departing from home and when leaving behind the *Everyday Space*. They remain in the *Tourist Space* while being on the road until after their return home.

Local Space

The *Local Space* corresponds to the *Everyday Space* from the perspective of a resident in a place visited by travelers. Since most travelers move along similar routes and make use of a tourist infrastructure, they are confronted with the *Local Space* only to a limited extent. Thus the *Tourist Space* takes up only a small part of the *Local Space* but the traveler enters and leaves both spaces at the same time.

In relation to the ten steps of a journey it may be summarized that the *Everyday Space* is as prevalent as the *Tourist Space* and the *Local Space* in contrast to a conventional approach to traveling focusing only on the time which is spent on the road.

Having determined the geographical spaces, I will next delineate the mental spaces.

Imagined Space

The *Imagined Space* includes the collective images and imaginations regarding a certain journey as well as the traveler's personal knowledge, expectations, role models, fantasies, and dreams. The *Imagined Space* plays an important part before, during, and after the actual journey.

Experienced Space

The *Experienced Space* covers the effective experiences and impressions of the traveler perceived on the road. It starts with the departure from

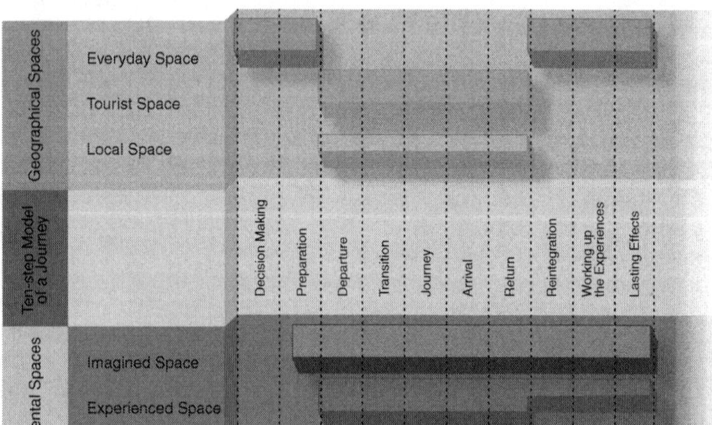

Figure 13.3 Graph, Multiple Spaces in the Course of a Journey.

the *Everyday Space* and blends with the *Imagined Space* immediately. Both continue simultaneously until the traveler's return when the *Experienced Space* remains commingled but in another form, namely through memories, records, and photographs.

Looking now at all the spaces in the course of a journey of ten steps it becomes evident that the *Imagined Space* takes up the most part of the entire journey and that it is not only relevant for the perceptions within the *Experienced Space* combined with the *Tourist* and *Local Spaces* but also thereafter in the *Everyday Space* of the aftermath. The graph (see figure 13.3) will help to visualize these thoughts and to put them across simply.

Conclusion

Coming home at last
At the end of the year,
I wept to find
My old umbilical cord.

(Bashō: *The Records of a Travel-worn Satchel*)[41]

Based on the theoretical structure of a journey consisting of ten steps plus empirical studies undertaken among travelers in Japan, Spain, Thailand, Cambodia, and Laos I have tried to demonstrate how a combination of multiple spaces is responsible for the perceived experiences of a traveler and how geographic spaces interfere with mental spaces and vice versa in the course of a journey. Starting from the assumption that the actual experiences a traveler senses during a journey merges with collective images and personal associations brought along, I elaborated a theoretical framework of the different geographic and mental spaces a traveler is confronted with. These spaces include the *Everyday Space*, the *Tourist Space*, the *Local Space* on the geographic side and the *Imagined Space* and the *Experienced Space* on the mental side.

During the last three steps of a journey many travelers describe their trip in clichés looking at their photos taken in "fictive spaces" as Hennig would call the blend of the *Imagined* and the *Experienced Space*.[42] As a result their accounts trigger the *Imagined Space* of their traveling successors on top of the latter's already existing collective images created by different media inputs—hence it's a never-ending move along a circular course.

For the empirical study of modern travelers I consider the distinction of multiple spaces in the frame of the ten-step model as a journey's structure especially useful, particularly when examining the experiences of travelers and their resulting transformation process. Distinguishing the different inherent spaces may better facilitate a researcher to sift out an informant's true feelings and experiences expressed in his/her statements from the media inputs and the hearsay that the informant was confronted with beforehand.

Last but not least, the proposed theoretical framework might also be valuable for researchers working empirically, particularly in order critically to scrutinize their own individual as well as collective images and the notions that constitute their hypothesis before going into the field, plus how their scientific observations correspond with their *Imagined Space* and with their quest of the already known.

Notes

1. Alex Garland, *The Beach* (London: Penguin, 2007 [1997]), 194.
2. See Jana Binder, *Globality: Eine Ethnographie über Backpacker* (Münster: Lit Verlag, 2005), 145–46; Camille Caprioglio O'Rielly, "From Drifter to Gap Year Tourist. Mainstreaming Backpacker Travel," *Annals of Tourism Research* 33 (2006): 1007–13; Tommi Mendel, "Foot-Pilgrims and Backpackers. Contemporary Ways of Traveling," in *Pilgrimage Today,* Tore Ahlbäck and Björn Dahla, eds. (Åbo: Donner Institute for Research in Religious and Cultural History, 2010), 299–301; Günter Spreitzhofer, "Drifting und Traveling: 30 Jahre Rucksacktourismus in Südostasien: (K)ein Beitrag zur Entwicklung? Eine Bilanz," in *Tourismus in der "Dritten Welt". Zur Diskussion einer Entwicklungsperspektive,* Herbert Baumhackl et al. eds. (Wien Promedia, 2006), 107–12; Michael Stausberg, *Religion im Modernen Tourismus* (Berlin: Insel, 2010), 15–16.
3. See Tommi Mendel, ARUKIHENRO: WALKING PILGRIMS, Ethnographic Documentary Film, Tigertoda Productions, (Watertown MA: Documentary Educational Resources, 2006), DVD; Tommi Mendel, "Arukihenro: Reflections on Fieldwork by a Visual Anthropologist," *Electronic Journal of Contemporary Japanese Studies* (2008): Discussion Paper No. 6, accessed October 30, 2011, http://www.japanesestudies.org.uk/discussionpapers/2008/Mendel.html; Mendel, "Foot-Pilgrims." Although the label of "pilgrim" is still generally associated with devout persons leaving home for purely religious motives, most protagonists in Japan and Spain doing the pilgrimages on foot are not religious in the sense that they feel close or attracted by a religious institution or its declared belief systems. In this article I deliberately do not distinguish between religious associated and secular associated forms of traveling.

4. The perceived images and experiences may be constructed for travelers as religious representations, yet they do not necessarily need to be so.
5. Hape Kerkeling, *Ich bin dann mal weg: Meine Reise auf dem Jakobsweg* (München: Piper, 2006), 105. Author's translation.
6. See Regina Scheyvens, "Backpacker Tourism and Third World Development," *Annals of Tourism Research* 29 (2002): 151; Peter Welk "The Beaten Track: Anti-Tourism as an Element of Backpacker Identity Construction," in *The Global Nomad: Backpacker Travel in Theory and Practice,* Greg Richards and Julie Wilson, eds. (Clevedon, UK: Channel View Publications, 2009), 80.
7. See next section.
8. Mendel, "Foot-Pilgrims," 289–90.
9. See Ian Reader and Tony Walter, eds., *Pilgrimage in Popular Culture* (New York: Palgrave, 1993); Simon Coleman and John Eade, eds., *Reframing Pilgrimage. Cultures in Motion* (New York: Routledge, 2004); Jill Dubisch and Michael Winkelmann, eds., *Pilgrimage and Healing* (Tucson: University of Arizona Press, 2005); Tore Ahlbäck and Björn Dahla, eds., *Pilgrimage Today* (Åbo: Donner Institute for Research in Religious and Cultural History, 2010).
10. See Tommi Mendel, "Kleidung und Körpererfahrung als Faktoren von Transitory Traveling Communities: Temporäre Identitäsbildung unter Fusspilgern in Japan und Spanien," in *Second Skin: Kleidung und Körper als Repräsentation von Religion,* Monika Glavac et al. eds. (Göttingen: Vandenhoeck & Ruprecht, 2013), 56–57.
11. Jack Kerouac, *On the Road: The Original Scroll* (New York: Penguin, 1957), 376–77.
12. See, for example, Binder, *Globality*; Nancy Louise Frey, *Pilgrim Stories: On and off the Road to Santiago* (Berkeley: University of California Press, 1998). Greg Richards and Julie Wilson state that the departure before the trip and the process of the return home had not been studied in any backpacker literature. Greg Richards and Julie Wilson, eds., *The Global Nomad: Backpacker Travel in Theory and Practice* (Bristol: Channel View Publications, 2009), 271, 278.
13. Barbara Haab, *Weg und Wandlung* (Freiburg: Universitätsverlag Freiburg Schweiz, 1998), 11; Klaus Westerhausen, *Beyond the Beach: An Ethnography of Modern Travelers in Asia* (Bangkok: White Lotus Press, 2002), 111.
14. Victor Turner and Edith L. B. Turner, *Image and Pilgrimage in Christian Culture* (New York: Columbia University Press, 1978), 249.
15. See Mendel, "Foot-Pilgrims," 295–96.
16. Turner and Turner, *Image and Pilgrimage*, 253.
17. As mentioned in note 4, for some travelers religious representations will occur as part of their perceptions in the journey, though certainly not for everyone.

18. Claude Lévi-Strauss, *Traurige Tropen* (Köln: Suhrkamp, 1978), 31. Author's translation.
19. Christoph Hennig, *Reiselust, Toursimus, Tourismus und Urlaub* (Frankfurt a.M.: Suhrkamp, 1999).
20. Hennig, *Reiselust, Toursimus*, 57–58.
21. Christoph Becker, *Paul Gaugin—Tahiti* (Stuttgart: Verlag Gerd Hatje, 1998), 37.
22. Becker, *Paul Gaugin*, 37.
23. Jutta Bertram, "Frohe Feste voll überschäumender Lebensfreude: Die Rolle der Exotik im Dritte-Welt-Tourismus," in *Trouble in Paradise. Tourismus in die Dritte Welt*, Christian Stock, ed. (Düsseldorf: Verlag Haus Altenberg, 1997), 102.
24. China Williams et al., *Southeast Asia on a Shoestring. Big Trips on Small Budgets*, 15th ed. (London: Lonely Planet Publications, 2010), 359.
25. Julie Wilson and Greg Richards, "Backpacker Icons: Influential Literary 'Nomads' in the Formation of Backpacker Identities," in *The Global Nomad: Backpacker Travel in Theory and Practice*, Greg Richards and Julie Wilson, eds. (Bristol: Channel View Publications, 2009), 123.
26. Welk, "The Beaten Track," 85.
27. The interview was conducted in Spain in 2009. Translation from Spanish to English by author.
28. The interview was conducted in Spain in 2009. Translation from Spanish to English by author.
29. The interview was conducted in Laos in 2011.
30. Bertram, "Frohe Feste," 108.
31. The interview was conducted in Switzerland in 2009. Translation from Swiss German to English by author.
32. Hennig, *Reiselust, Toursimus*, 38, 56.
33. Hennig, *Reiselust, Toursimus*, 94–95.
34. The interview was conducted in Spain in 2009. Translation from German to English by author.
35. The interview was conducted in Spain in 2009. Translation from German to English by author.
36. Raimund Joos and Michael Kasper, *Outdoor: Der Weg ist das Ziel. Spanien: Jakobsweg. Camino Francés* (Welver: Conrad Stein, 2009), 147. Author's translation.
37. Jörg Schader, for example, writes: "Especially in the Meseta, in the barren plateau between Burgos and Léon, spiritual experiences get piled up" (author's translation). Jörg Schader, "Jakobspilgern: Eine teilnehmende Beobachtung am Camino Francés" (PhD diss., Karl-Franzens-University Graz, 2009), 149.
38. Binder, *Globality*, 48.
39. Hennig, *Reiselust, Toursimus*, 96. Author's translation.

40. William Sutcliffe, *Are You experienced?* (London: Penguin, 1997), 226–27.
41. Matsuo Bashō, *The Narrow Road to the Deep North and other Travel Sketches* (London: Penguin, 1966), 77.
42. See note 17.

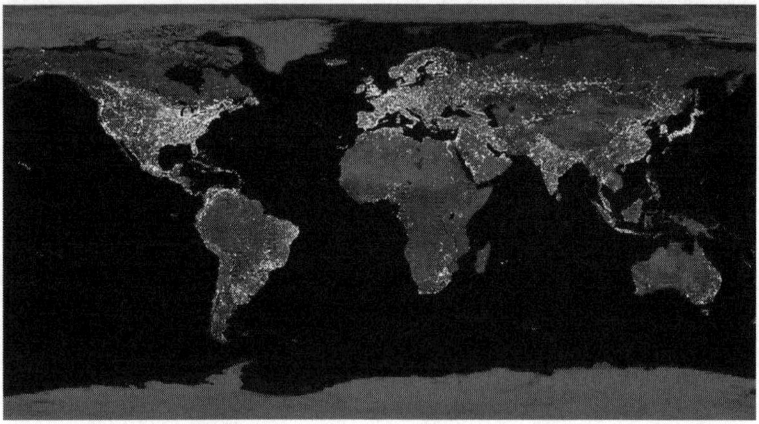

Figure 14.1 The Earth by night, photographed from a NASA-satellite.
Source: WikiMediaCommons.

Chapter 14

The Town beyond the Modern: A Biographical Reflection about Space, Time, and Spirituality

Carl Fingerhuth

Drawing on my own career experience, in this chapter I outline changes in urban gestalt, space, and time[1] (see figure 14.1). In the last 50 years from Europe to Africa to China, cities have moved beyond the modern age and into a new reality where Western rationality is transcended by emotion and spirituality. We are now in the midst of a radical transformation of consciousness. One of the many changes brought on by this shift is a new awareness of the collective self, of the physical body as incarnation of the self, and of the soul as part of our cosmic origin. This also informs a shift in how we perceive our cities: they are no longer viewed as simple piles of rocks but as society's physical "body" innately connected to our collective knowledge and cosmic source. From within a framework of gestalt therapy as a starting point, the gestalt of the city with its new energies is connected to emotion, sensitivity, and spirituality in today's world. The goal is to overcome the fundament.

Introduction

Working with towns and their space so that they become an intense reflection of their inhabitants' needs, goals, and dreams—in other words, creating and maintaining meaningful town space—has been my lifelong preoccupation. Towns are where I gained experience, learned, and developed. In my professional life, from 1956 until today, I have had the opportunity to experience from within the framework of towns this dramatic change in human consciousness as we transition from modernity into the current era. I came of age just as the modern dogma was coming into question. As such, I had to deal with, both inwardly and outwardly, today's new, dramatic and yet unnamed age. My best guess is to call our time the era beyond the modern. I believe that my background as an architect makes me especially aware of the transformation we are experiencing. To use the world of the town to draw attention to this shift makes a lot of sense: like art, religion, or politics, the town is one of the most intense reflections of our being, our aspirations, and our past. Perhaps because the world of the town is so close to us, we do not acknowledge this meaningfulness.

A Biography as a History Book

My first town-related assignment was in 1958, working in the *Office pour la construction des tours de Carouge* in Geneva. Next to the Carouge historic center, five *Machines à Habiter* were to be built according to the rules of the "Athens Charta," Le Corbusier's post–World War II treatise that at the time had come to represent ultimate town-planning dogma (see figure 14.2). The project clearly demonstrated the consciousness of the time: functionality, rationality, purity, and transparency were its key issues. This meant a clear separation of housing, work, traffic, and recreation, ignoring the existing town, and celebrating "form follows function" and "less is more."

Figure 14.2 I started by being a modern architect in Geneva. Reproduced by permission of Carl Fingerhuth.[2]

For my diploma in architecture in 1960, I was asked to design a technology museum with the stipulation that each department within the museum could be extended or reduced independently. There was no specific site; it should function as a prototype that could travel all over the world (see figure 14.3).

My quest for a more complex view of towns began with a job as an archaeologist in Egypt. We were charged with finding remnants of

Figure 14.3 A modern museum.

Figure 14.4 I went back to the past by working as an archaeologist in Egypt.

past cultures before they were lost due to the construction of the new high dam at Aswan (see figure 14.4). Our primary focus was a temple built during the rule of Ramesses II. During the site's first excavation in 1920, the early Coptic Church, situated in the courtyard of the temple, was deemed a nuisance and torn out without anyone recording its existence.

The extraordinary Nubian villages, doomed to disappear under the water of the new high dam, met the same fate. Archeology at that time meant deciding which layer was the relevant one and removing anything that might detract from the splendor of the "royal" culture (see figure 14.5).

Upon my return to Switzerland, I was not sure where to go. So I declared myself an expert in town and country planning. This was a new and unknown field, open and undefined, which at the same time was attracting growing public awareness.

My first job involved establishing an agency for town and country planning in the state of Wallis, a very conservative world that harbored a lot of skepticism toward new ideas (see figure 14.6). A primary issue of contention concerned creating zoning laws in order to be able to connect sewage to a water treatment station. But almost overnight, everything changed. The old times were gone and we suddenly faced catastrophe: a typhoid epidemic broke out in the famous ski resort of

Figure 14.5 Kalabsha in the Nubian desert.

Figure 14.6 Verbier Switzerland, 1965.

Figure 14.7 Master plan for the area of Bremgarten in the metropolitan area of Zurich, 1972.

Zermatt. People died because of water pollution. I was confronted with one of the first enigmatic signs of the upcoming environmental crisis.

When I turned 28, I decided to end my "apprenticeship" and open my own town planning and urban design company. I focused on bringing elements of space together (housing, public buildings,

Figure 14.8 Presentation of the master plan for the capital of Imo State Nigeria, 1976.

Figure 14.9 The center of Basel, Switzerland.

traffic, landscape) and developing new ways to engage people in public planning processes (see figure 14.7).

This period ended with a contract to design the capital of a newly created state in Nigeria (see figure 14.8). The governor of the state told me he wanted a city like Paris or London. We tried to look for African identity; it was a fascinating and a very conflictual assignment!

At 42, my time as an "expert" was finished. I was offered a position as state architect for the city of Basel, Switzerland. I was no longer a consultant. I was given political, economic, and social responsibility and a lot of money to invest in the transformation of a 2000-year-old town (see figure 14.9).

I stepped right into the middle of a cultural and political battle. On one side were the old planners, staunch supporters of the modern town vision with its high-rise buildings, private cars, parking houses, and urban highways right through the center of the historic city. On the other side a new generation of citizens who opposed any change at all of the existing town (see figures 14.10a and b).

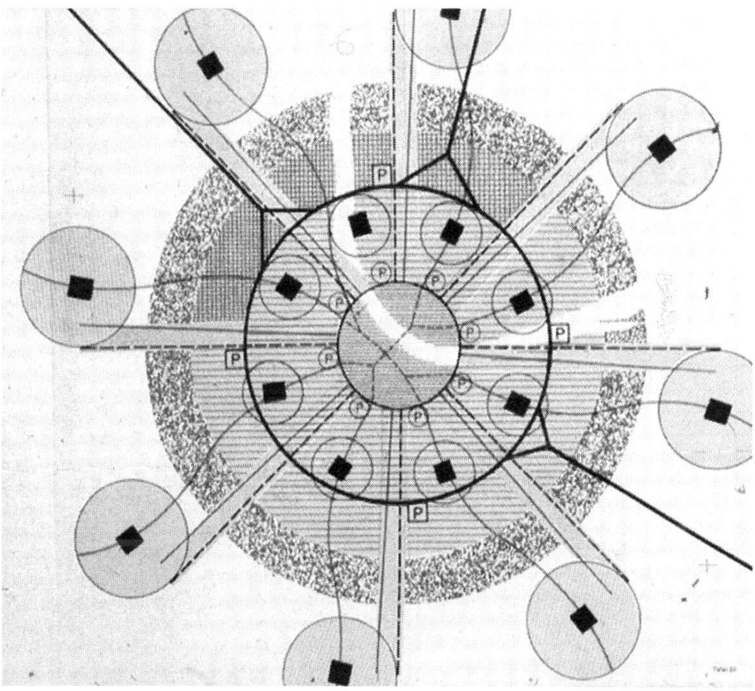

Figure 14.10a Master plan for Basel, 1960.

Figure 14.10b Election poster of the green party "Social instead of concrete," 1980.

Finding consensus, bringing together form and structure (or gestalt) to represent a town beyond the modern, was a fascinating job. The project I developed on the Hebelstrasse with the architects Herzog and de Meuron became a worldwide icon (see figure 14.11).

In 1992, after 14 years in politics and administration, I returned to a life of personal independence. I am now, once more, a consultant

Figure 14.11 Wooden house at Hebelstrasse in the center of Basel.

Figure 14.12 The author talking about Basel.

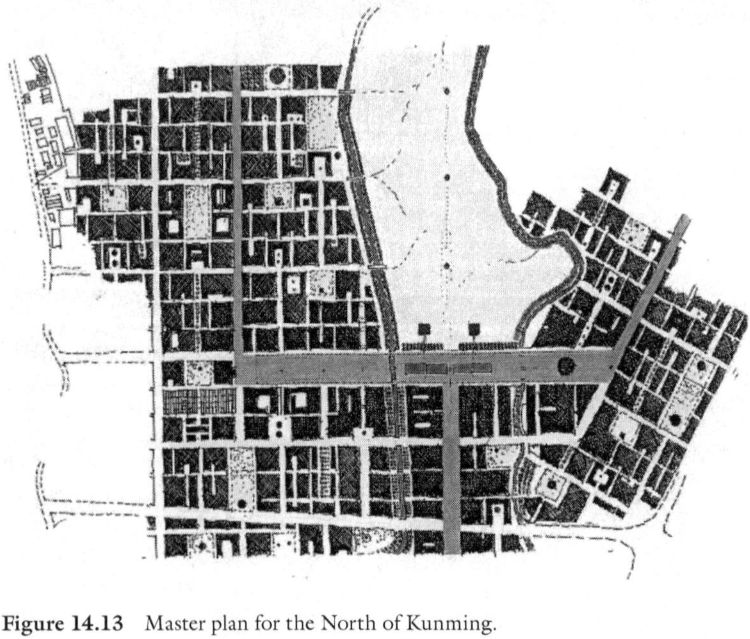

Figure 14.13 Master plan for the North of Kunming.

for towns and I share my experience by teaching at universities and writing about the town (see figure 14.12).

In this process I became more and more drawn in the interaction and interdependence between the new emotional and spiritual world of our time and the physical manifestations of towns. It appeared in discussions about Feng-Shui lines in Kunming, China, or when I was invited to join discussions about space and spirituality at universities (see figure 14.13). The question was, how to develop new forms of dialog in processes connected with taking care of the transformation of towns.

Understanding the Town

Towns Are Picture Books

Towns tell in one picture one hundred stories, each one with ten thousand words. Words are hard and often one-dimensional. Pictures are open and multidimensional. (If I were permitted to rewrite the bible, I would propose changing the wording of John 1:1 from: "In the beginning was the word," to: "In the beginning was the picture.") Therefore, towns do not use words but show themselves and tell their story with pictures, with gestalt—form and structure with a meaning.

Towers in Thailand, in Switzerland, in the United States: all three have a square ground floor, all three dominate their surroundings, but each one tells a different story (see figures 14.14a, b, c).

The stupa in Thailand is a spiritual symbol. It reminds man of the unity of Earth and Heaven. On the way up to the sky the square form of the ground floor, as a symbol of the physical, transforms itself into a circle, as a symbol of universal unity, which then dissolves itself into the form of a pointed needle in heaven into pure spirituality.

This Basel high-rise apartment scoffs at nineteenth-century city housing blocks and shows off the boundless potential of the modern world. The inherent promise is that modern technology will eliminate all boundaries and social inequalities. It promises a glorious new town that will replace the dark old city and its restrictions.

The World Trade Center in Manhattan was the ultimate symbol of the importance and dominance of the economic power of Western civilization, reducing the Statue of Liberty to a puppet, and becoming the main physical focus point of the whole of Manhattan, indeed, of the Western world. It did not make friends and celebrated through its gestalt the power of rationality and perfection. Moreover, it wasn't

Figure 14.14a A stupa in Thailand.

Figure 14.14b A tower in Basel.

Figure 14.14c Two towers in the United States.

Figure 14.15 Mexcaltitan, Mexico, probably the first capital city of the Aztecs.

wearing a "crown" like the Chrysler building, but was prepared to continue growing right up into the sky.

The structure of Mexcaltitan could be seen as a reflection of the inner world of the Aztec civilization (see figure 14.15). The universal circle was structured by an orthogonal order, a first symbol of human beings controlling the world. The intersection of Broadway, part of an old Indian path connecting the Hudson River to the ocean, and Fifth Avenue, is a silent reference to the roots of New York.

Both pictures prove our entry into a new era. We are in the midst of a radical transformation of consciousness. One of the many changes this shift entails is the perception of our collective self, from the physical body as incarnation of the self, to our souls as part of our cosmic origin. This produces a similar shift in how we perceive our towns. They are no longer regarded as simple piles of stones. Thus, towns are thought of as the "body" of society, connected to its collective consciousness and its cosmic origin.

Where Are We and Our Towns and Where Are We Going?

Bookstore display windows are filled with books about our inner and outer world. There are books about the ego, morphogenetic fields, polar ice caps, the difference between Islam and Christianity. The subject of towns is rarely integrated into any of these books. There are books about towns but mostly they are picture books, monographs, and guidebooks. They show images of the world or praise the latest hero of architecture.

There is a fascinating coherence of phenomena between mankind, town, and the cosmos. Phenomena could be defined as "a fact or situation that is observed to exist or happen, especially one whose cause or explanation is in question."[3] These particular phenomena are not new, but they were buried, or discredited, because they were a nuisance in the context of the dogma of the modern era.

My approach is different; gestalt is my frame of reference. I seek to connect the gestalt of towns to the new energies which show themselves: the emotionality, sensitivity, and spirituality of our world today. Bridging the great divide between society and the technicians of our towns will allow cities to develop according to the needs, wishes, and dreams of their inhabitants. The term "town" refers to space that has been transformed by human intervention. It includes not only the densely built-up areas of the common towns but also all the transformations of nature inflicted by modern civilization: ski resorts in the Alps, shopping centers and gas stations along highways, or electrical high-tension lines and pipelines.

"Cosmos" refers not only to the physical universe but also to mankind's inner world. Also, I use the word "mankind" in a multidimensional way. It stands for the global and the individual. Today's change of consciousness is a global one, with ramifications across many different cultures. But at the same time there is a strong reaction to globalization, which reactivates specific local energies.

Common Phenomena of Mankind, Town, and Cosmos

The common phenomena of mankind, town, and cosmos are all organized in holonistic structures. Every element is part of a more complex element and contains in itself less complex elements. A house is not a house if it doesn't contain rooms and a town is not a town if it doesn't contain houses.

They all require structure and form for their organization and identity. Language is ruled by grammar and orthography. The grammar of a town is defined by its systems: the sequence of its open spaces, public transportation stations, or the main pipelines of the sewage system. Houses and gardens are the words. They are all a palimpsest with old, irrelevant text scraped off, replaced by new text. The old text is mostly imperceptible, but it is still there.

Awareness of the polarity between all things was essential for all pre-modern philosophies. The era beyond the modern reintegrates this knowledge. It is no longer a matter of right or wrong, but a consideration of less or more. Too much change in a town creates a feeling of insecurity. Not enough change leads to a feeling of decay and hopelessness. Taking care for the transformation of the town in the time beyond the modern means balancing these two polar energies. Given the space limitations of this publication, I will only explore the subject of evolution, both in theory and practice integrating the new beyond the modern paradigm.

From Mythical Space to the Town beyond the Modern

Evolution

The Polish philosopher Jean Gebser's description of the steps of evolution—from archaic, to magical, to mental, and eventually to what he called integral consciousness—has become a standard interpretation of human evolution and works wonderfully to understand towns[4] (see figure 14.16).

Gebser uses different "structures" to explain the steps of evolution. He begins by looking at our consciousness of space. Mankind's understanding of space has evolved from a nonexistent awareness of spatiality to a four-dimensional consciousness. I eschew Gebser's term "integral" for the current age. It sounds too final. There is no evidence that evolution will not go on. Five-dimensional consciousness is next; one day, we will have to find a term for that era! As a society, we

Structure	Dimensioning	Perspectivity	Emphasis	Settlement
Archaic:	Zero-dimensional	None Pre-temporal	Prespatial	Cave
Magic:	One-dimensional	Pre-perspectival Timeless	Spaceless	Village
Mythical:	Two-dimensional	Un-perspectival Natural temporality	Spaceless	Town
Mental:	Three-dimensional	Perspectival Abstractly temporal	Spatial	Agglomeration
Integral:	Four-dimensional	Aper-spectival Time-free	Space-free	Global City

Figure 14.16 Space and time relationship according to Jean Gebser; Settlement column added by the author.

do not yet have a fitting term for the current era. We used to define it as postmodern. Unaware of how autonomous and powerful these changes would become, architects discredited the word. Therefore I prefer to characterize our time as the era beyond the modern.

The only physical clues we have of the zero-dimensional space in archaic time are some footprints in the East African soil.

Very sophisticated paintings of animals, painted 13–14,000 years ago appeared out of nowhere. Mankind started to perceive the world and memorize its gestalt. Human beings began creating places in space.

About 4,000 years ago, settlements appeared around the world, structured along lines corresponding to astronomical information. In India, China, Italy, and Mexico town layouts transformed from organic sprawl into circles, then into rationally defined squares. These mythic-based structures change over thousands of years into a new consciousness. Abstract temporality and spatial thinking characterize the new mental era. It develops worldwide and embraces the third dimension of space. Our perception of Planet Earth morphs from flat to globe shaped. Towns begin building towers and artists discover perspective.

But evolution didn't stop. With Albert Einstein's theory of relativity, the limitations of three-dimensional space disappeared and were replaced by a time-space continuum. Space is now seen as a dynamic system. Einstein's drawing with a person on top of a train and a person standing next to the train refers to the connection of time and space. The two persons will perceive the flashlight atop the speeding train at different times. If the train is running at the speed of light, the observer on the train will never see it. Simultaneousness is relative. Any concern about the transformation of towns beyond the modern must integrate this change from a static perception of space to a dynamic perception of space.

About the Origins of Separation

Preparing a speech at the Japanese-German Institute in Berlin, I found another fascinating approach to the topic of evolution. In 1948 the Japanese philosopher Watsuji Tetsuro wrote about the interaction between climate and religion.[5] He described religion as a product of mankind's trust in nature.

Christianity has its origin in the climate of the desert. Nature was hostile. Humans will not be fed if they do not work: "In the sweat of your face you shall eat bread till you return to the ground" (Gen.

3:19). Humans had to create their own urban paradise, with walls to protect them from nature—its physical and spiritual dangers. They built churches like fortresses and towns with stone walls around them.

The religion of the north Asian, European, and American continents had its roots in what Tetsuro called the meadow climate. Nature was friendly. Humans were sure that in spring the apples and nuts would again grow for them. They were not chased out of paradise, so they did not need walls to protect themselves. The holy place was in nature, under the tree. As they were part of all, they did not need to separate themselves from nature. Buddhism was a product of the monsoon-climate. Nature provided food and water in abundance. But you could not trust. Periodically nature became so destructive that everything was washed and blown away. The settlements on earth were insecure and provisional. The best solution was trying to find your way to Nirvana, a "place of oblivion to care, pain or external reality."[6] So there was no reason to invest in a solid town.

In our time beyond the modern, religions are gradually replaced by a global spirituality. In this consciousness nature is basically not regional but global. Climate is seen more and more as an indication of the earth's situation. Importance is given to the status of the seas, the forests, the overall temperature, or the ozone shield. Everything is interdependent. Global nature provides the energy, the sun, the wind, the tides, and the rain. I am fascinated by the presence of these different cultural backgrounds in the north European discussions about urban design and town planning.

Understanding the Town beyond the Modern

The Time beyond the Modern

The needs, values, and longings of our world are changing dramatically. Our consciousness is more complex. We must integrate the new realities into the gestalt of our towns and find the appropriate form and structure. We must give form to the formless (see figure 14.17).

One of the major problems of our time is that in architecture, urban design, and town planning, politicians and professionals hesitate to acknowledge or even seek to discredit this paradigm shift—the new set of values that define our collective thinking, feeling, and sensing today.

The modern era was an expedition into the depths of our thought potential. This obsession was so strong that other potentials—feeling,

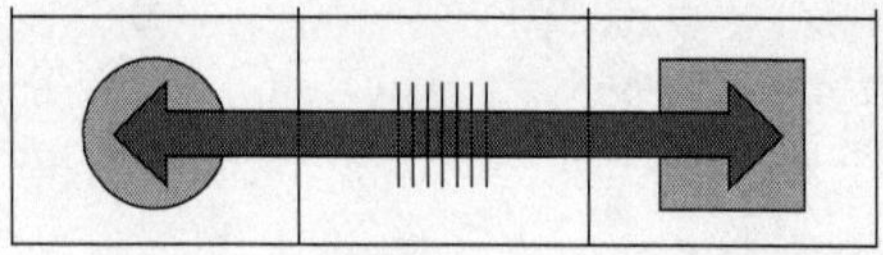

Society	Building	Space
Needs	Planning	Plot
Goals	Urban Design	Area
Values	Architecture	Town

Figure 14.17 Giving form to the formless.

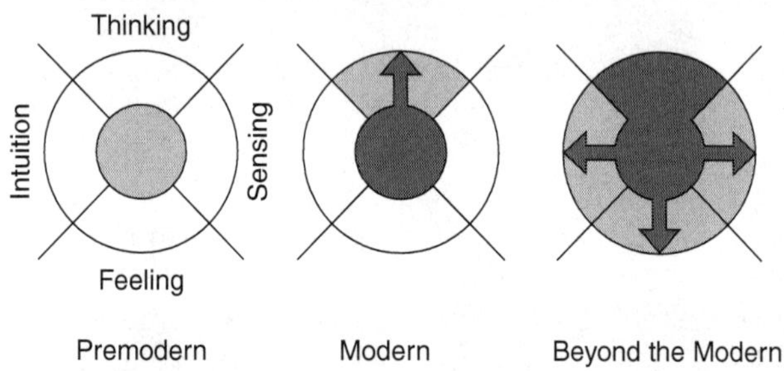

Figure 14.18 Expansion of human consciousness.

sensing, and intuition—were neglected or even aggressively discriminated against. Today, more and more phenomena appear that point to the overbearing focus on our rational potential of the modern time. Strong energies appear that draw our attention to this default. The scientific world is scared by these phenomena. I will try to point out some of the main themes of this new quest (see figure 14.18). Due to limitations of space, these themes will only be spotlights on a vast territory. C. G. Jung over and over again stressed the importance of this assignment: "I can only take the reaction which begins in the West against the intellect in favor of Eros, and in favor of intuition, as a mark of cultural advance, a widening of consciousness beyond the too narrow limits set by a tyrannical intellect."[7]

The Longing for Reintegration of Feeling and Sensuality

Our collective yearning for the reintegration of feeling and sensuality has a tremendous impact on how we deal with our towns. One of the huge challenges of our time is to qualify this longing, to transform

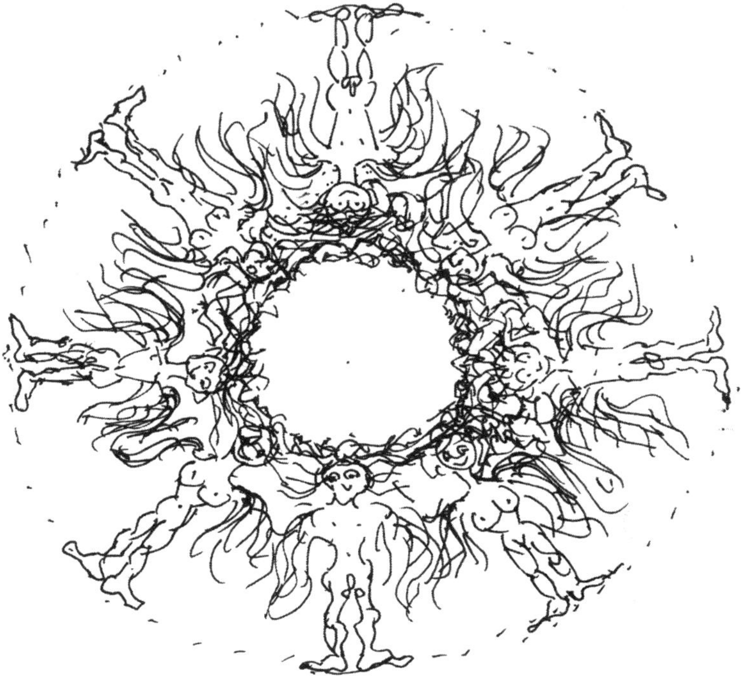

Figure 14.19 Drawing by the author.

Note: This drawing was published in Cordula Rau, ed., *Sehnsucht: The Book of Architectural Longings* (Wien: Springer Vienna Architecture, 2010) for the German Pavilion at the Architecture Biennale 2010 in Venice.

spatial "pornography" and coarseness into sophisticated and loving sensuality (see figure 14.19).

In 1972 the Austrian painter Friedensreich Hundertwasser started painting houses as wonderful colorful fantasies. Then he proclaimed that the painters had to build the houses, because according to him beautiful houses had disappeared. The title of an exhibition he had in 2005 in the German Museum for Architecture in Frankfurt was "Built Dreams and Longings." In a review of the exposition, journalist Dieter Bartetzko wrote: "Irritating like a dilettante but accurate like a caricature, Hundertwasser's orgies of Kitsch denounce the lacks of today's architecture. One does not have to take the buildings seriously but reflect on their popularity."[8] In 1977 the city of Vienna asked Hundertwasser to build a house on the Löwengasse. In the first years it was visited by more tourists than the castle of Schönbrunn.

The Longing for Reintegration of Spirituality

Spirituality no longer represents the opposite of materialism. Instead it symbolizes a new consciousness of the unity of all things, a dream that embraces all aspects of life. As such it becomes a comprehensive term for all the potentials of mankind. This newly emerging spirituality is independent of all religions. Religions are not anymore dogmas but paths to a spiritual consciousness of the holiness of all things.

In this understanding the town is not any more just a pile of stones. Accepting the concept of the holonic structure, it is a reflection of ourself on a comprehensive level. We have to take care of the town in the same way as we take care of ourselves.

The Longing for the Unity of Man and Nature

In his book *No Boundaries*, Ken Wilber describes the specific theme of separation in modern society: separation of persona and shadow, body and soul, mankind and nature.[9] Our new spirituality feels different. We long for union on all levels and this we need also in our towns.

The town beyond the modern faces a comparable challenge: apartments crave to have loggias, balconies, or terraces, houses would like to have gardens, and towns long for trees and parks.

The New Challenge

Care for towns beyond the modern is not radically different from how we cared for modern towns. Clean water and sewage are still delivered and removed in the same way. What is changing is our consciousness. As we confront new energies, we are developing:

- an awareness of the polarity of all things,
- an awareness of the complexity and contradiction within towns,
- an awareness of the imprints of past cultures within our towns,
- an awareness of the importance of reintegrating the potentials of the self (sensuality, emotionality, spirituality) discarded by modernism, and
- a need to heal our separation from nature.

Architects must stop complaining that they are no longer heroic creators. They have to accept that they are midwives, translators, transformers working for the benefit of the population. We are moving toward an awareness without dogmas, where people accept the complexity and contradiction of our world, where they want to understand what is going on in Cairo, Stuttgart, or Beijing, and where they want to be respected and not exploited. This will require a new gestalt in the form and structure of our towns, and it will call for new "software," new instruments, methods, and procedures, as we are entering a new era beyond the modern.

Meaningful towns are aesthetic not anesthetic; they are beautiful not banal; and they are in accordance with the needs, values, and longings of people. When they are all these things, they will also be spiritual towns. This is one of the big challenges of our time.

Notes

1. This contribution is based on a presentation given at the international conference "Meaningful Spaces" in April 2011 at the University of Zurich. One different version of the text has been published in: Christoph Woiwoode and Wolfgang Schotz, "Urban Development and Spirituality Trialog," *Zeitschrift für das Planen und Bauen in der Dritten Welt* 107.4 (2010): 18–26.
2. Unless otherwise indicated, all subsequent figures are reproduced by permission of Carl Fingerhuth.
3. *Oxford Dictionaries Online*, s.v. "phenomenon," accessed April 12, 2013, http://oxforddictionaries.com/definition/english/phenomenon.
4. Jean Gebser, *The Ever-Present Origin* (Athens: Ohio University Press, 1985).
5. Tetsuro Watsuji, *Fudo–Wind und Erde: der Zusammenhang zwischen Klima und Kultur* (Darmstadt: Wissenschaft Buchgesellschaft 1997).
6. *Webster's Ninth New Collegiate Dictionary*, s.v. "Nirvana."
7. C. G. Jung and Richard Wilhelm, eds., *Geheimnis der Goldenen Blüte* (München: Diederichs, 1990), 64.
8. Dieter Bartetzko, "Alarmstufe Kitsch," *Frankfurter Allgemeine Zeitung*, November 29, 2005.
9. Ken Wilber, *No Boundury: Eastern and Western Approaches to Personal Growth* (Los Angeles: Center Publications, 1979).

Chapter 15

The Ambiguities of Preservation: Bodhgayā, UNESCO, and the Making of a World Heritage Site

Jacob Kinnard

Who's Temple Is It?

The Mahabodhi Temple Complex has outstanding universal importance as it is one of the most revered and sanctified places in the world.

—Information Dossier for nomination of Mahabodhi Temple Complex as a World Heritage Site.[1]

On June 30, 2002, dozens of Buddhist monks walked in a procession through and around the Mahābodhi Vihāra at Bodhgayā. They were celebrating the announcement, three days earlier, of the designation of Bodhgayā as a World Heritage Site (WHS) by the United Nations Educational, Scientific, and Cultural Organization (UNESCO). The WHS designation was hailed as a great triumph for Buddhists throughout the world; it was a recognition, by the world's most prestigious arbiter of history and culture, of the significance of Bodhgayā, placing the Mahābodhi Vihāra among the ranks of over 800 other World Heritage Sites, including the Taj Mahal, Angkor Wat, Macchu Picchu, and the Great Wall of China. The monks paraded around the temple and town with huge banners, attracting a small crowd of followers. Although it was a rather sober sort of celebration, to anyone who had been in and around Bodhgayā consistently over the prior decade, this celebration would have stood in sharp contrast to many of the processions at the temple in the 1990s: a series of protests and conflicts over control and management of the temple complex, which, although never escalating to actual physical violence, always carried with them the threat of conflagration. In the weeks and months after

UNESCO's designation, as the implications of this honor began to emerge, questions and anxieties began to surface. What did all this mean, really? Who and what would be affected by the designation? What would complying with the various WHS rules entail? Rumors spread that the shops and tea stalls and dwellings surrounding the temple complex would need to be removed; some worried that the many monasteries that had sprung up over the last century would be demolished. Was this really, in the end, a positive development?

On its face, the designation of Bodhgayā as a WHS represents a significant commitment by UNESCO—with its nearly half-billion-dollar annual budget—and the government of India to help preserve this most important of Buddhist sites. But we must ask: Why was this designation sought in the first place? What does it mean to preserve such a temple? Who decides the mechanisms of such preservation? Unlike the vast majority of WHS sites in India (there are, at this writing, 27 such sites), Bodhgayā is not simply an ancient architectural structure; it is a living temple. It is not a museum, but an active and fluid place, one that has over the past 1,500 years been constantly changing—often as a result of contestation over what sort of place it is. It has been repaired and, essentially, rebuilt numerous times; it has been abandoned at various points in its history; old structures have decayed and been demolished; new structures have been added. It has been both a Buddhist and a Hindu place of worship.

Thus as much as the WHS designation can be celebrated because it seems to ensure that the temple will be preserved, it is not clear at all that this preservation will be for the Buddhists and Hindus who actually participate in the life of the temple. The WHS was designed to preserve what are understood, explicitly, to be places of "outstanding universal value" that are "part of the world heritage of mankind as a whole."[2] These are places that "belong to all the peoples of the world, irrespective of the territory on which they are located." But what does this mean, to say that places such as the Mahābodhi Vihāra "belong" to all the people of the world? Why should UNESCO dictate how an Indian Buddhist temple is managed? Is this yet another example of Western hegemonic discourse and practice, simply colonialism dressed up in the new clothes of cultural globalization? I certainly do not wish to suggest that monuments such as Bodhgayā should not be preserved; rather, I wish to interrogate the processes and politics of such preservation. What I will argue here is that the plan outlined for the Mahābodhi represents, really, the "museumizing" of the temple complex, which, far from presenting a neutral program of restoration and preservation, instead poses a real threat to the life of the temple, and is itself as much about transformation as about preservation.

The Application for Inclusion

The "Convention Concerning the Protection of the World Cultural and Natural Heritage" that was adopted by the United Nations in 1972 is UNESCO's root preservationist text, presenting the organization's guiding vision for why preservation of the world's cultural monuments is necessary. It is a fascinating document, a kind of head-on collision between postcolonial inclusivism and proto-globalization arrogance. The *leit motif* of the document is the phrase "outstanding universal value," a phrase that occurs throughout the sixteen-page text. The central concern of the 1972 convention is the creation of an "Intergovernmental Committee for the Protection of the Cultural and Natural Heritage of Outstanding Universal Value, called 'the World Heritage Committee,'" a body of, initially, fifteen states—subsequently expanded to forty—that is charged with collectively identifying and protecting those places in the world that, according to UNESCO, are of "outstanding interest and therefore need to be preserved as part of the world heritage of mankind as a whole."[3] In other words, these are sites that may be owned, literally, by a particular group or a particular state, but which are really the property of the whole world.

The convention is intended, fundamentally, to protect and preserve those sites of "universal value," but it is not, despite the list, clear what UNESCO is protecting the sites from. There are the obvious natural dangers—serious fires, earthquakes, landslides, volcanic eruptions. There are also less obvious dangers, though, less obvious in the sense that some of these dangers are, in fact, inherent in the very nature of places such as Bodhgayā. The convention mentions "large-scale public or private projects of rapid urban or tourist development projects; destruction caused by changes in the use of ownership of the land; major alterations due to unknown causes; abandonment for any reason whatsoever." Are all of these in fact dangers?

Let me begin with the first here, "large-scale public or private projects of rapid urban or tourist development projects." When the Government of India drew up its revised application for WHS status for Bodhgayā, it turned to Aswini Lohani, the chairman of the Indian Tourism Development Corporation. In 1999 he had been the driving force in gaining WHS designation for the Darjeeling Himalayan Railway, the seventeenth cultural site in India to be declared a WHS by UNESCO. Although there is no question that both of these sites are of cultural and historic significance, the applications for both put a marked emphasis on their potential as tourist attraction; as such, the tourism potential of Bodhgayā and the infrastructure to support it would clearly need to be substantially developed.

The "Justification for Inscription" section of the application begins by stating that Bodhgayā is a "sanctified" place of universal importance.[4] The "Justification" then goes on to describe the cultural and historic significance of the Mahābodhi Temple; it quotes Faxien and Xuanzang, the fourth- and seventh-century Chinese pilgrims who described their visits to Bodhgayā; it quotes the *Mahāparinibbāna Sutta* of the Pāli Canon; it quotes Alexander Cunningham, the director of the Archaeological Survey of India who oversaw the restoration of the site in the 1880s; and it quotes several contemporary Indian archaeologists. The dossier recounts the history of the temple in considerable detail. For the most part, there is nothing surprising or controversial about this historical narrative, although it is almost exclusively presented from a Buddhist perspective, and as we have seen, Buddhists are not the only group with a history at (and of) the temple, or who have made claims on the temple.

One of the key parts of the revised WHS dossier is a twenty-page "Management Plan," which, in response to the rejection of the original application, outlines in great detail the specific management, conservation, and development of the site. One of the unusual things about the Mahābodhi temple is that it is a site that already had a management structure in place, the Bodhgayā Temple Management Committee (BTMC), which was formed in 1949 in response to the decades of wrangling over who should have control of the temple.[5] The WHS application simply states that the BTMC will continue to make all management decisions regarding the temple complex. The Conservation Plan and Development Plan, however, present rather more complex and ambiguous documents, elaborating on suggestions that were made prior to the submission of the first WHS application, in 1999, by the Patna Circle of the Archaelogical Survey of India, at the request of the BTMC.

"Fixing" Bodhgayā

The conservation plan seems, on its face, to be a straightforward document, offering up a list of needed repairs: "Votive stupas in the parapet wall need to be reset and re-plastered so that they do not fall off. Vegetation growth on the Temple structure needs to be removed effectively."[6] Drainage needs to be improved, plaster needs to be restored, the grounds tidying. None of this is particularly controversial at first glance. Underlying these seemingly innocuous repairs, however, is a notion of what the Temple "originally" was and the state to which it should be restored. For instance, some plastering that was done in a 1953–1954 restoration needs to be replaced with "original materials." "Acryclic emulsion paint has also been used in some place on the sculptural figures. This changes the original character and beauty of the figures and needs to be removed."[7]

These seemingly neutral observations echo, albeit with a more neutral tone, those of Sir Edwin Arnold from a century earlier; the author of *The Light of* Asia had found Bodhgayā to be in a deplorable state, and he complained, as we have seen, that the Hindus who had been using the temple since it was abandoned by the Buddhists in the thirteenth century had defaced "the shrine with emblems and rituals foreign to its nature."[8] Clearly, though, to the Hindus who venerated Buddha images as Viṣṇu, or who worshipped *stūpas* as Śiva *liṅgams*, such rituals were hardly "foreign" to the objects to which they were directed. The underlying assumption here is that physical objects such as temples and sculptural images (as well as the people who venerate them) are static entities with fixed identities. Bodhgayā, like most temples in India—and in the world—has been in an almost perpetual state of restoration; it is nearly impossible to determine what the "original" temple might have been or what it should be. Furthermore, this quest for the "original" is a distinctly Western endeavor, one that cannot easily, or at least not unproblematically, be imposed on the Indian context.[9] As Tzvetan Todorov has quipped, "only dead cultures remain intact."[10] Nonetheless, the Conservation and Development Plan articulated in the WHS application explicitly sought to "fix" the temple, in at least two senses of the word: to repair it, but also to make it permanent. However, to do so was also to make significant changes to both the physical and ritual life of the temple, a paradox that goes unmarked in the WHS application, the consequences of which would continue to surface at Bodghayā years after the WHS designation.

The use of acrylic paint and other "non-original" repairs do not evoke the same antipathy as Hindus transforming what was once a Buddhist *stūpa* and venerating it as a Śaiva *liṅgam*, but not all of the Conservation Plan is concerned with such seemingly inconsequential matters. The plan suggests that some of the stones and epigraphs located in the floor of the temple need to be removed to a separate gallery; no reason is given, but presumably they risk damage. There is no mention of how long these objects have been in situ, or what their function is:

> Devotees have been following the practice of lighting oil lamps or candles along all the walls and railings and on sculpted figures in the Temple complex. This is adversely affecting the ancient monument and spoiling the pathways. The burning of a large number of oil lamps on festive occasions is a threat to the structure of the Temple as well as to the Bodhi Tree. A thorough cleaning of the oily residue on all parts of the Temple needs to be carried out and an alternative found to this practice of burning oil lamps.[11]

It is the last part of this statement that is particularly striking, since it involves not simply restoring the physical structure of the temple, but altering the ritual practices that are an integral part of its life as an active religious place. This comes up in the main body of the application, where it is stated that this ritual practice "needs very much to be controlled." The document suggests that the "alternative" would be to create a special place for the ritual, out of the way, an alternative that would protect the temple "without offending the sentiments of the devotees." There are, the document goes on to say, "many places where the Buddhist *Rinpoches* have been persuaded to take an enlightened view to discontinue the burning of oil lamps near the place of worship. Instead, they have located a place well removed from the shrine where a single oil lamp is kept burning with the offerings made by devotees."[12] The Conservation Plan elaborates what the alternative would be: the BTMC "has earmarked a different site south of the Temple where it intends to build a glass house with a modern exhaust system."[13]

There is, certainly, a clear recognition that the Mahābodhi temple complex is a dynamic, living space, and that any archaeological work on the site must attend to the competing understandings and uses of the space. That said, however, the historic consistently trumps the contemporary. Restoration and preservation of the "original" temple is the primary goal of the archaeological work to be done. The contemporary is recognized as important to the life of the temple complex, but in a quite limited way: "But the growing dimensions of building construction whose visual density goes to mar the beauty of historic colour, looks wild . . . Modern construction . . . should be in accordance with the principle of harmonic integration in which modern building will be designed in the same spirit as the contemporary traditional buildings."[14] This points to what is perhaps the key tension in restoring a place like Bodhgayā, a tension that is certainly not unique to this context.[15] If one were to read just this section of the Information Dossier, one might come away sanguine that the competing understandings of the temple complex—historic/original versus contemporary/living—could be balanced. The emphasis on the "original," however, trumps the more dynamic understanding of the Mahābodhi as a changing, fluid, living place.

There are at least two visions of Bodhgayā that emerge from the WHS application, two overlapping, interrelated visions, to be certain, but also two visions that stand in tension with one another. The first is the "outstanding universal value" vision that, essentially, would freeze Bodhgayā; or, more accurately, would restore Bodhgayā to what is was originally and then freeze it. I think of this as the "fixing" vision: the temple would be "fixed up," and then it would be "fixed" in time. It would, essentially, be museumized. The second vision is the "living temple" vision. This is, really, a kind of background vision, a rather reluctant recognition that Bodhgayā is not simply a culturally and historically significant place to visit, but a place where devout Buddhists and Hindus go to perform what they understand to be essential religious rituals.[16]

The lines between tourism and religious pilgrimage have, no doubt, always been blurred; indeed, the very attempt to make a sharp distinction, an either or distinction, seems misguided—although scholars often attempt to do precisely that.[17] This is not to say, however, that the two modes are without tensions; in the WHS application, there is a marked tension between the desire to preserve/restore the "original" Mahābodhi Vihāra and the desire to promote it as a tourist destination, a tension that the document's authors are at least aware of: "The approach adopted is to prepare a holistic development plan with clearly identified projects including clear implementation strategies. The focus of the development plan will be tourism promotion, while conserving the local environment and the local resources."[18] The BTMC and the Housing and Urban Development Corporation of India (HUDCO) proposed to build a meditation park, a museum of Buddhism, and reception and information centers. Furthermore, the WHS application notes that Bodhgayā is very hot in the summer, which "makes it very difficult for tourists to enjoy the experience of walking around the temple and benefiting from the peaceful atmosphere of the site." It would, then, be "desirable to plant an adequate number of trees" around the temple "which would surely effect the environment positively."[19]

Such developments seem harmless enough. Others, like the proposal to move the ritual lamps to a separate area, are more ambiguous, and seem to ignore or at least to discount the ritual and religious aspects of the *vihāra*. For instance, in the Development Plan appended to the application, a great deal of attention is paid to cleaning up and beautifying the temple grounds. "Innumerable monuments, stupas and memorials built by visitors and dignitaries to the temple during past centuries are located haphazardly and needs re-orientation...it will be better if the smaller 'stupas' could be re-arranged along the pathways or in the form of Stupa Garden, with proper grills and placards giving its history, if possible."[20] Although the many *stūpas* surrounding the main temple may "clutter" the area, to many Buddhists they are integral to the ritual life of the temple; others serve as reliquaries for significant monks. Their "haphazard" arrangement reflects centuries of use, layers and layers of ritual practice.

Later in the Development Plan, it is proposed that the various small *stūpas* be rearranged in a systematic manner:

> This of course can be done if it is approved by the religious leaders.... It is further recommended that all the remains be identified, through the experts, so that their historical importance could be established. Once this is done, the brief may be placed in form of a placard next to each object so that the worshipers and the visitors could realize the importance of each of the remains. In due course an album for reference could also be prepared.[21]

Knowingly or not, the authors of the WHS application, following the lead of the BTMC and HUDCO, are here replicating the language and *modus operandi* of their former colonizers. Among the various ways the British sought to control India was by collecting and organizing physical objects, particularly art and artifacts. As Bernard Cohn has pointed out, the eighteenth- and nineteenth-century colonial administrators in British-occupied India sought to control—and just as often to fabricate—the history of India in part by controlling the physical objects that were the products of that history.[22]

Transforming Bodhgayā

It may be that what we see in the WHS application is a kind of postcolonial taxonomic hangover that necessitates clear categorical divisions of function and space—Hindus here, Buddhists there; religious activities here, tourist activities there. It may also be that the authors of the application simply recognized—and capitulated to—the orderly, "modern," "scientific," Western, European orientation of an organization such as UNESCO.[23] It is, at any rate, worth recalling that one of the fundamental purposes of the 1972 "Convention Concerning the Protection of the World Cultural and Natural Heritage" was, as the title so clearly denotes, to protect endangered cultural sites, and one of the explicit threats to such sites, as I have noted, is tourist development.

One of the difficulties for the authors of the Bodhgayā WHS application was to come up with a plan for both development and conservation that would also recognize the polyvalent nature of Bodhgayā, or at least the polyvalent nature as it was conceived by those seeking the WHS designation. This is an attitude that is, certainly, in sharp contrast with the colonial taxonomic ideology that saw temples and objects as only possibly belonging to a single group. To many Buddhists, however, the Mahābodhi *vihāra* could be nothing but a Buddhist temple. Indeed in the years after the WHS designation, this would become a decidedly heated issue. At any rate, although I have named two different conceptions of the Mahābodhi complex that can be teased out of the Information Dossier, the authors of the WHS dossier overtly adopted an essentially binary approach:

> In view of the outcome of the analysis and physiographic character of the site, the entire Bodh Gaya Temple Complex Development has been broadly conceived as a two way functional approach as follows:
>
> 1. Meditiation-Religious
> 2. Tourist Attraction.
>
> The two aspects of the design are entirely diverse and do not go hand in hand. A religious person wants peace to worship for prayers and to listen to the religious discourses whereas a tourist visits the Temple to have glimpses, pay his regards as well as enjoy the time that he is in the complex.[24]

The problem, according to the authors of the Development Plan, is that, at the time of the application, the two aspects were not distinct in the temple complex, and "the result is chaos."[25]

They offered up a plan that would substantially transform the space and therefore the way in which that space could be used by those who claim it, a transformation that, as we shall see below, had serious unintended consequences. It separates the two groups, moving the various religious "functions"—meditation platforms, sculptures, *stūpas*—out of the way of the tourists. What the plan fails to recognize or acknowledge is that there are, as I have already noted, no clear distinctions between religious visitors and pilgrims; it is a false distinction, one that is more informed by the post-Enlightenment West's distinction between the secular and sacred than it is by the reality played out on the ground at a place such as Bodhgayā. Furthermore, what does it mean to say that the temple complex is chaotic? What is the implied (normative) model of order?

To strip away what was "added," as the British routinely did, was to radically alter—really, it was to do violence to—the character of the temple. As Byrne, writing about the Thai context, has noted, the archaeologists' zeal for the "original" is, ultimately, misguided in the Buddhist context: "The ethic of merit making in Thai Buddhism, moreover, puts a premium on the proliferation, reconstruction and rebuilding of *stupas*—practices which routinely imperil or consume the original fabric and structure of old and often ancient *stupas*."[26] Interestingly, although his analysis is very much critical of the sorts of "fixing" that leads to Foucault's "ceremony of objectification," Byrne slips into the very language of those who he is critiquing here—"routinely imperil" the "original." Indeed, we are so used to thinking in these terms that the very language we use is bound by the ideology of preservation.

What Byrne describes is not true only in the Thai context: the history of Indian temples is fundamentally a fluid history, a history of images being altered or "converted" from one tradition to another, of the fluidity of objects and materials. Western methods of preservation and conservation have tended to ignore such material fluidity, however; the effects of such methods are on display in virtually any art museum that contains Hindu or Buddhist sculptures.[27] Such images are transformed into art, and as such bear almost no resemblance to what they would have been in their temple settings.

Although the WHS dossier recognizes that the Mahābodhi is a living temple, and although UNESCO—in their various reviews of the application and subsequent assessments of the implication of the Development Plan—seems to recognize the various religious groups who have a stake in the temple, the very structure of the WHS designation precludes any real engagement with the complex issues on the ground. Bruce Owens has examined a similar situation at Swayambhu, in Nepal:

> The struggles now being waged over cement, stone, and turf at places like Swayambhu are often struggles over whether cultural practices or their material consequences should be privileged as worthy of preservation, and determining whose culture, as of when, is at stake for

> whom. Thus distinguishing between new and old kinds of transformations at these places is more than an exercise in distinguishing postmodernity from modernity, or modernity from whatever came before. These distinctions lie at the heart of debates taking place between those determining what these places will look like in the future, or even whether or not they will continue to exist.[28]

These, however, are precisely the complex issues that the authors of the WHS dossier for Bodhgayā do not consider, and, really, precisely the tensions that UNESCO strategically does not engage.

The WHS designation reflects an essentially monothetic, static conception of Bodhgayā; the Mahavihara is, according to this conception, an archaeological site, an ancient monument that must be restored to its original state and preserved for the whole world. In this sense, then, UNESCO becomes the ultimate arbiter of what is and what is not legitimate at Bodhgayā; in so doing, however, it stands outside of the fray, apparently uninvolved in the mess on the ground, objective, rational, and scientific, in much the same way that the British Orientalists did in the eighteenth and nineteenth centuries: "Eighteenth and early nineteenth-century epistemology was thus rooted," writes Edney, "in a vision which, with its surrogates, established an almost physical distance between the viewer and the viewed, between the subject and the object of vision."[29] In a study of the eighteenth and nineteenth centuries' genre of landscape painting in Britain, Ann Bermingham has argued that the landscape painter presents "an illusionary account of the real landscape while alluding to the actual conditions existing in it."[30] This is relevant to the colonial context, of course, because the observers of India were informed by this same basic ideology. Much the same sort of "illusory account" can be seen in Bodhgayā's WHS dossier as well, an account endorsed, after all, by UNESCO. But it is important to recognize that, much like the "objective" view presented in British landscape painting, "the claim to possess a realist and naturalistic view is itself an ideological statement of power."[31] Again, we see the West, as embodied by UNESCO, holding the Indian other at a distance just as it controls it with its discourse of order, rationality, and science. As the subaltern project has demonstrated, the irony is that this discourse has been internalized;[32] in order to receive the approval of UNESCO, the Indian applicants for the WHS designation necessarily had to adopt the assumptions of UNESCO, and, as we have seen, they did so using the vision of India—religion, culture, art—constructed by the Orientalists, an illusionary, idealized, static vision.

One important element of this WHS program of preservation is the creation and maintenance of the "buffer zone," the area where only those structures designated as "original" can be located. Accordingly, the revised master plan submitted to UNESCO by HUDCO dictates that all shops and commercial complexes within the 1-km buffer be relocated. In early 2008, a group made up of members of a diverse collection of business organizations, independent traders, hotel associations, and travel agencies marched in protest in Bodhgayā because the "clean up" of the buffer zone would imperil their livelihoods. Here, again, we see a clash of visions about what the temple complex should be and, what is more, the ironic inconsistency of any single vision of the temple.

Conclusion

It is difficult to see how the different religious groups can reconcile their very different visions of the temple complex. And what of the merchants and traders? Although they are not part of UNESCO's vision of the temple—which of course simply assumes the Enlightenment separation of the sacred and the secular—such people and the various activities in which they are engaged have always been an integral part of the life of Indian temples. As Vincent Berdoulay has pointed out, "a place comes explicitly into being in the discourse of its inhabitants."[33] Ironically enough, the master plan itself also depends on them; a substantial component of this vision of Bodhgayā is commercial. As with the candles and "new" images and all of the other "additions" to the Mahābodhi complex, however, it seems that they must be put in their proper place—they must be ordered and controlled—outside of the temple complex itself.

One of the consequences of UNESCO's ideology may be, then, that places such as Bodhgayā will, for all intents and purposes, be killed in order to be preserved. In an interesting article on the great temple at Borobodur, Shelley Errington has adopted the framing language of Irving Goffman to analyze the different ways in which the government of Indonesia, UNESCO, and others created a particular vision of Borobodur that, in significant ways, has become its own on-the-ground reality. These various agents, and the government of Indonesia and UNESCO in particular, have created a kind of "totalizing" frame for Borobudur—which was declared a WHS in 1991—thereby "eliminating other 'frames,' other 'lives," other stories about it."[34] Errington picks up Appadurai's notion of the "social life of things,"[35] and suggests that understanding physical objects—including monuments and temples—in this way "has the virtue of encouraging us to notice that an object's meaning is not located in a single moment of its history" and that what "an object is may change in the course of time, as it is moved from one place to another and changes hands and contexts."[36] Significantly, Errington was told by one person whom he interviewed that after Borobodur had become a WHS, it had "lost its soul" and was, essentially, dead.[37]

At Bodhgayā, the Buddhists see the Hindus, and the Indian government that is constituted by them, as the enemy; the Hindus, in turn, see the Buddhists, with their exclusive and exclusionary demands on the Mahābodhi, as the enemy. UNESCO, however, remains remote and neutral and, therefore, blameless. The issue, in the end, is not whether a place like Bodhgayā will be transformed; certainly it will, as it has continuously been transformed since its beginnings. Rather, the question is who will dictate the terms of this ongoing transformation.

Notes

1. General Conference of the United Nations Educational, Scientific and Cultural Organisation, "Information Dossier for nomination of Mahabodhi Temple Complex, Bodhgaya as a World Heritage Site," 4, accessed July 1, 2013, http://whc.unesco.org/uploads/nominations/1056rev.pdf.
2. General Conference of the United Nations Educational, Scientific and Cultural Organisation, "Convention Concerning the Protection of the World Cultural and Natural Heritage," Paris, November 16, 1972, 1, accessed July 1, 2013, http://whc.unesco.org/en/conventiontext/.
3. "Convention Concerning the Protection of the World Cultural and Natural Heritage," 1; a second document specifically addressing nonmaterial aspects of culture—aspects as diverse as the watertight-bulkhead technology of Chinese junks to Cambodia's Royal Ballet to Indonesian puppet theater and French gastronomic meal. General Conference of the United Nations Educational, Scientific and Cultural Organisation, "Convention for the Safeguarding of Intangible Cultural Heritage," Paris, October 2003, accessed July 1, 2013, http://www.unesco.org/culture/ich/en/convention.
4. "Information Dossier, Management Plan Mahabodhi Temple Complex, Bodh Gaya," 4.
5. See Alan Trevithick, *The Revival of Buddhist Pilgrimage at Bodh Gaya (1811–1949): Anagarika Dharmapala and the Mahabodhi Temple* (Delhi: Motilal Banarsidass, 2006), and Tara Doyle, "Bodh Gayā: Journeys to the Diamond Throne and the Feet of Gayāsura" (PhD diss., Harvard University, 1997).

6. "Information Dossier, Management Plan Mahabodhi Temple Complex, Bodh Gaya," 3.
7. "Information Dossier," 3.
8. Sir Edwin Arnold, *East and West, Being Papers Reprinted from the "Daily Telegraph" and Other Sources* (London: Longmans, Green and Company, 1896), 310.
9. That said, however, it has been a typical *modus operandi* in Buddhist Studies since the discipline began, with numerous claims to uncover the "true," "original," "primitive," and so on teachings of the Buddha. For a specifically archaeological treatment of this issue, see Denis Byrne, "Buddhist Stupa and Thai Social Practice," *World Archaeology* 27.2 (1985): 266–81.
10. Tzvetan Todorov, "Tribunals, Apologies, Reparations, and the Search for Justine: In Search of Lost Crime," *The New Republic* (January 2001): 9–10 (9).
11. "Information Dossier, Management Plan Mahabodhi Temple Complex, Bodh Gaya," 3.
12. "Information Dossier," 27.
13. "Information Dossier," 4.
14. "Information Dossier," 12.
15. See J. O. Blatti, ed., *Past Meets Present: Essays about Historic Interpretation and Public Audiences* (Washington: Smithsonian Institution Press, 1987); Jean Misitelli, "World Heritage, between Universalism and Globalization," *International Journal of Cultural Property* 11.2 (2002): 323–36. For UNESCO, see Thomas Erickson, "Between Universalism and Relativism: A Critique of the UNESCO Concept of Culture," in *Culture and Rights: Anthropological Perspectives,* Jane Cowan, ed. (Cambridge: Cambridge University Press, 2001), 127–48.
16. There are, really, two "subvisions" here: one is a Buddhist vision of the temple, one a Hindu vision. But even this is too simple, as we shall see below, because different Buddhists communities have different visions of the temple complex.
17. See, for example, Erik Cohen, "Pilgrimage and Tourism: Convergence and Divergence," in *Sacred Journeys: The Anthropology of Pilgrimage,* Alan Morinis, ed. (Westport: Greenwood Press, 1992). Cohen takes a structuralist approach, and argues—following Turner—that there is a basic difference, in that pilgrims travel to a center of the world whereas tourists travel from a center to a periphery. Cohen critiques those scholars who have seen a modern/premodern distinction, or a material/spiritual distinction, but in the end I find his own binary analysis altogether too neat.
18. "Information Dossier, Management Plan Mahabodhi Temple Complex, Bodh Gaya," 21.
19. "Information Dossier," 27.

20. "Information Dossier," 7.
21. "Information Dossier," 11.
22. Bernard S. Cohn, *Colonialism and Its Forms of Knowledge: The British in India* (Princeton, NJ: Princeton University Press, 1996), 302, 304. See also James Clifford, "On Collecting Art and Culture," in *The Predicament of Culture: Twentieth Century Ethnography, Literature, and Art,* James Clifford, ed. (Cambridge, MA: Harvard University Press, 1988), 215–51. For more on the taxonomic tenor of British Orientalists in India, see Cohn, *Colonialism and Its Forms of Knowledge: The British in India* (Princeton, NJ: Princeton University Press, 1996), and Nicholas B. Dirks, *Castes of Mind: Colonialism and the Making of Modern India* (Princeton, NJ: Princeton University Press, 2001); see also Tim Barringer and Tom Flynn, eds., *Colonialism and the Object: Empire, Material Culture and the Museum* (London: Routledge, 2004), and Tony Bennet, *Pasts Beyond Memory: Evolution, Museums, Colonialism* (London: Routledge, 2004).
23. For a formative view of UNESCO, see Julian Huxley (the first director of UNESCO), *UNESCO: Its Purpose and Philosophy* (Washington: American Council on Public Affairs, 1947); see also Thomas Ericksen, "Between Universalism and Relativism: A Critique of the UNESCO Concept of Culture," in *Culture and Rights: Anthropological Perspectives,* Jane Cowan, ed. (Cambridge: Cambridge University, 2001), 127–48; Jean Musitelli, "World Heritage, Between Universalism and Globalization," *International Journal of Cultural Property* 11.2 (2002): 323–36; Robert Shephard, "UNESCO and the Politics of Cultural Heritage in Tibet," *Journal of Contemporary Asia* 36.2 (2006): 243–57.
24. "Information Dossier, Management Plan Mahabodhi Temple Complex, Bodh Gaya," 9.
25. "Information Dossier," 9.
26. Byrne, "Buddhist Stupa and Thai Social Practice," 267.
27. There is, perhaps, reason to hope that there has been something of a shift in the practices of museum collection and display; see M. Simpson, *Making Representations: Museums in the Post-Colonial Era* (London: Routledge, 1996).
28. Bruce McCoy Owens, "Monumentality, Identity, and the State: Local Practice, World Heritage, and Heterotopia at Swayambhu, Nepa," *Anthropological Quarterly* 75.2 (Spring, 2002): 269–316 (271).
29. Matthew Edney, *Mapping an Empire: The Geographical Construction of British India, 1765–1843* (Chicago: University of Chicago Press, 1999), 48.
30. Ann Berminham, *Landscape and Ideology: The English Rustic Tradition, 1740–1860* (Berkeley: University of California Press, 1996), 3.
31. Edney, *Mapping an Empire*, 57.

32. The literature on this topic is extensive; see, for example, the varied essays in *Subaltern Studies X*, Gautam Badhra, Gyan Prakash, and Susie Tharu, eds. (New Delhi: Oxford University Press, 1999); Dipesh Chakrabarty, *Habitations of Modernity: Essays in the Wake of Subaltern Studies* (Chicago: University of Chicago Press 2002), Vinayak Chaturvedi, ed., *Mapping Subaltern Studies and the Postcolonial* (Verso: London, 2000), and David Ludden, ed., *Reading Subaltern Studies: Critical History, Contested Meaning and the Globalization of South Asia* (London: Anthem South Asian Studies 2002).
33. Vincent Berdoulay, "Place, Meaning, and Discourse in French Language Geography," in *The Power of Place,* John A. Agnew and James S. Duncan, eds. (London: Unwin Hyman, 1989), 124–39 (135).
34. Shelly Errington, "Making Progress on Borobudur: An Old Monument in New Order," *Visual Anthropology Review* 9.2 (1993): 32–59.
35. Arjun Appadurai, *The Social Life of Things: Commodities in Cultural Perspec*tive (New York: Cambridge University Press, 1986).
36. Errington, "Making Progress on Borobudur," 32; see also Richard Davis, *The Lives of Indian Images* (Princeton, NJ: Princeton University Press, 1997).
37. Errington, "Making Progress on Borobudur," 56.

Conclusion

The Space of Interpretation: A Hermeneutical Rereading

Pierre Bühler

Introduction

This book contains a broad set of contributions coming from very different disciplines. Even if the main accent lies on the study of religion (seven articles) and, connected with it, on biblical studies (Hebrew Bible, Old Testament), the volume also includes developmental pediatrics, astrophysics, geography, philosophy, cultural theory, and architecture. When the topic is the study of religion, the variety of approaches also is very rich: visual media like painting, ornament, caricature, fiction film, but also clothing, pilgrimage, or preservation of sites, which open many new perspectives. This large panel of contributions creates a *space* of interdisciplinarity that makes the reading of the book very interesting. As an interdisciplinary challenge, the topic of "meaningful spaces" is a hermeneutical one: the space of interdisciplinarity is, in that sense, *a space of interpretation*, a space in which different interpretations of space may meet, not for a great synthesis—that would be quite boring!—but for creative interactions. In this same sense, Paul Ricoeur spoke of a fruitful *conflict of interpretations*.[1]

The Importance of Space and Its Inherent Tensions

Such creative interactions bring awareness to the importance of the topic of space and its multiple facets, which are revealed in the course of the book. Space belongs fundamentally to the human condition; in other words, there is no human existence without its spatial concretization, realized in many different ways. Even if Martin Heidegger's *Being and Time*, as the title suggests, underlines the dimension of time, there is also an important accent on the "In-der-Welt-Sein," the "being in the world," as a fundamental spatial condition of human existence. This is even more stressed in Maurice Merleau-Ponty's *Phenomenology of Perception*: a whole chapter is devoted both to the reality of space in the perception of the world and to the spatiality of one's own body.[2]

The central signification of space for the human condition also has to do with its difficulties, its tensions. In a very superficial manner, we could think of space as something evident and simple, and often we probably live in a spontaneous way in space without being aware of its complexity. We just *use* it, as a domesticated, insignificant reality, presupposing its availability, disposing of it as something that does not matter. But if we take it into account in a deeper way, we may suddenly realize how *inherent tensions* let it be an earnest challenge. Because the meaning of space is never just given, it has always to be found, to be conquered again.

As pointed out from the beginning of the book, space has contrasted dimensions: it can be a very small world, for example the small world of a child learning to cope with spatiality in its first perceptions of its environment and its first actions in it, as the pediatrician Remo Largo notices. But in the widest sense, it also can be the seemingly infinite cosmos of the astrophysicist, as Arnold D. Benz argues. Both belong to the human perception of space, in an irreducible tension, and thus there are people suffering psychologically on one side or the other: *agoraphobia*, if the space gets unbearably big, and *claustrophobia*, if it gets unbearably small. These conditions show how important is the relationship to space for the feeling of life.

In his *Pensées,* Blaise Pascal thinks about this double perspective on space, the infinitely small and the infinitely big.[3] If a human being looks at the wideness of the universe, he or she not only feels reduced to nothing, but also "lost" in this wideness. On the opposite end of the scale, if he or she looks at the infinitely small, he or she feels colossal, like a monster. Some decades later, in the year 1726, Jonathan Swift developed this double perspective in telling the story of Gulliver's travels, being a dwarf under giants and a giant under dwarves!

Therefore, according to Pascal, humankind has to live in a strange "middle between nothing and all":

> That is our real condition. That makes us unable to know with certainty and to ignore absolutely. We sail on a vast middle, always uncertain and floating, pushed from one end to the other. Every boundary where we were thinking of tying and strengthening ourselves gets loose and

> leaves us, and if we try to follow it, it escapes from our grips, slips away and flees in an eternal flight; nothing stays quiet for us.[4]

What a contrast with the radical and ultimate certainty of his contemporary Descartes! In the middle between two infinites, without any final certainty: we can understand that, for Pascal, as Arnold Benz quotes, space is something frightening. But maybe it is a fright every human being can experience when thinking about being in the world! Therefore we can understand all the attempts to get situated in space, to arrange it in conformity with our own wishes, as different ways to cope with this elementary and difficult uncertainty inhabiting our relationship with space. As Pascal adds in his thought, making a link with the tower of Babel: "This is the condition that is natural for us, but at the same time it is the most opposed to our inclination. We are consumed by the desire to find a firm situation, and a last constant basis for constructing on it a tower that rises up to the infinite, but our foundation cracks and the earth opens down to the abyss."[5]

Of course, Henri Lefebvre is right, as the contributions of the book often underline: space is not just given as a fact. It is also *produced* by different human mediations. In other words, it is a product of social, cultural, economic, and political processes constructing space, both for groups and for individuals. But at the same time, from a hermeneutical perspective, this production of space must be considered under the aspect of its *responsiveness*: it is an answer to the insecurity and uncertainty we feel in our relationship to the space that is *given* to us. Space is of course produced, but that is only a part of what has to be said. It would be wrong to reduce it just to this pure *activity*. The "producing" of space echoes the fundamental *passivity* given in the experience of space in its paradoxical dimension: as Pascal states, every boundary "gets loose and leaves us, and if we try to follow it, it escapes from our grips, slips away and flees in an eternal flight; nothing stays quiet for us." Because the space given to us can also be quite threatening, we work on it, creating space that suits us.

Therefore, we may interpret all manner of getting along with our space as a struggle in theory and praxis with its unseizableness. We are measuring, we are mapping, we are planning and preserving, we are arranging it. We represent it in paintings, drawings, movies. In a very global sense, we are dwelling in it, by thinking and acting, but finally also by feeling it, an aspect we maybe tend to forget because we are focused on our attempts to grasp it. When Merleau-Ponty describes the human perception of world, he starts with feeling (*sentir*), and that leads him to stress living in the world. He speaks of "lived space" (*espace vécu*).[6] That could be the main aim in all coping with space: living in it even if "nothing stays quiet for us" (Pascal), finding an existential certainty able to resist the constant uncertainty of the world we live in.

As the whole book underlines, *religion* also belongs, in strong association with living, to this effort of coping with space, in ways that are not completely different from those we just described, even if it does so in a specific manner. But for understanding that, we need to develop other dimensions.

Space in Movement

We could think of space as a static, fixed reality. But it would be a reductionistic idea. Because it is a "lived space," it also has to be perceived as a dynamic, moving reality, in many ways. As our book shows, space gets *represented*, through images, paintings, films, and drawings, and realized in different times, from the past and from contemporary contexts. These representations are not just different neutral descriptions of one unchanged reality. They draw perspectives on space, modeling it from different points of view. To use Clifford Geertz's distinction, the various representations of space are not only models *of* the world, they also are models *for* the world. Therefore space changes, gets new facets through its representations, and it is one of the strengths of this book to work with a number of images as examples of the visualization of space. They show space as a moving reality, in a kind of kaleidoscopic perspective that stimulates again and again the questioning about meaninglessness and meaningfulness. By their multiplicity, they contribute to the enrichment of what we called "the space of interpretation." But at the same time, space is also moving because there are bodies moving in it! That is another important topic of the book: the embodiment of space. As already mentioned, for Merleau-Ponty, corporeality is a central aspect of spatiality, and this connection appears at different places in the book: in ancient Israel's tabernacle and the English caricature of a Methodist worship service, in pilgrimage and the child's discovery of its own body's spatiality. These bodies are various figures of lived space.

From a theological perspective, we could say that this embodiment corresponds to the idea of incarnation. When the Gospel of John says: "The Word became flesh and made his dwelling among us" (1:14 NIV), it expresses how God's presence becomes a body in our space, setting it in a new light, making our space a place where humans may live freely.

The question of the meaninglessness or meaningfulness of space is not a question just hanging somewhere in the air or one to be resolved solely on paper. It is first an embodied question, a question which these bodies are coping with through their life in space. It is a lived question, and thus an open question, accompanying our sailing "on a vast middle, always uncertain and floating, pushed from one end to the other," as Pascal said. Representations, visions, words, and parables propose various models for answering this question, in a way giving us certainty in the uncertainty.

The book takes its readers along in the movements of these human bodies which are trying to find their ways in this moving space. The question of meaning is not an abstract question that can be examined in the aseptic climate of a laboratory. It is a concrete question. The space of interpretation is within lived space, and we question and answer by moving inside of its dynamics, working with figurations we can inherit from past and present testimonies.

Space and Exteriority

By sailing on the "vast middle" that is our real condition, we also get constantly to the boundaries, and it is an important experience to perceive space from its limits. Of course, it is not possible to jump over these limits and escape, allowing us to look at the limits from above or beyond them. It is just the possibility to get some distance for a moment, to realize momentarily space from outside, to get back to it with new intuitions, to look at it with a new perspective. That is what Ricoeur calls "the hermeneutical function of distanciation," which is important for every effort of interpretation.[7] Appropriation as the main goal of interpretation goes always through distanciation, or in the words of his reading theory: as a reader, I can only find myself by losing myself.

In that sense, there are various figures of distanciation through the book, giving to the readers moments of exteriority. A main one is the *viewer*, a topic appearing at several places in the book, for example as a Muslim looking through a window at a Methodist worship service! But this viewer also plays an important role in looking at cinematic spaces in journey movies or in discovering the mapping of apocalyptic spaces at Chartres. This viewer is a kind of hermeneutical subject, marking the process of interpretation. By identifying with him, the reader gets an outside perspective on one's lived space, allowing him to interpret it within the perspectives suggested by the book.

But the exploration of the boundaries of space from an outside perspective can happen in other ways, for example by looking at older representations of the otherworld, or by going as a pilgrim to unknown spaces. What took place as the embodiment of space in ancient Israel's tabernacle was also situated "outside the camp some distance away" (Exod. 33:7 NIV).

In a biblical perspective, we could say that this exteriority, allowing a person to cope with lived space by going for a moment beyond its boundaries, is a kind of a marker for divine transcendence. In Pascal's words: "These extremities [in space] touch each other, get together by dint of moving away from each other and meet each other in God, and in God only."[8] In this sense, questioning the meaninglessness and meaningfulness of our space leads to the question of God. Do spaces get meaningful by belonging to God's creation? Or by being the place where God decided to reveal himself?

There is an interesting dimension in biblical anthropology for our topic: human life is understood in a spatial sense by being placed "in front of God," "before God," or literally from the Hebrew, "before the face of God." Therefore the main question about human beings is not what they are or what they do, but *where they stand*. The relationship with God is the space of a vis-à-vis. Therefore, when Jacob left the place of his nightly fight with God in the morning, he called this "space of struggle" Peniel, which means in Hebrew, "Face-of-God." The place where the struggle took place is the space in which God showed his face. At the same time, this struggle was embodied in Jacob: "The sun rose above him as he passed Peniel and he was limping because of his hip" (Gen. 32:30–31 NIV).

Inspired by this aspect of biblical anthropology and its reuse in Luther's theology, Gerhard Ebeling speaks, in his theological hermeneutics, of the anthropological reality of "coram-relations," the Latin "coram" meaning "in front of": before God, but also before the world, before others, before oneself. Therefore Ebeling conceives human existence as a "forensic existence," from the Latin "forum," meaning the public place that was also the place for court trials. For our topic, that means human existence is a trial space where human beings are at stake, where they are questioned, accused, pardoned, condemned, and freed in front of the different instances of judgment.[9]

In this context, we could say that the questioning about the meaninglessness and meaningfulness of space is a constant trial in which every human being, as a responsible body in space, has to answer to the challenge of meaning. The meaning of the space in which we live is an existential task.

"Outside the Camp" (Heb. 13:11–14)

As a biblical example for interpreting the meaningfulness of space from an outside point of view, we can briefly evoke the Epistle to the Hebrews in the New Testament. As in ancient Israel the tabernacle is "outside the camp," so has Christ's death happened "outside the camp" for the author of this epistle. The parallelism is striking, making of Jesus a kind of new tabernacle: "The high priest carries the blood of animals into the Most Holy Place as a sin offering, but the bodies are burned outside the camp. And so Jesus also suffered outside the city gate to make the people holy through his own blood" (13:11–12 NIV). Jesus's crucifixion marks the divine exteriority in relation to the city as a structured, organized space. Jesus's crucifixion is an "outspace" offering a new perspective. Therefore the author deduces immediately an imperative concerning the relationship to the "normal" space: "Let us, then, go to him outside the camp, bearing the disgrace he bore" (13:13 NIV). The believers are inside the city no more: they are called to go "outside the camp," joining Jesus and his disgrace. The next verse expresses what this signifies for the meaning of the world: "For here we do not have an enduring city, but we are looking for the city that is to come" (13:14 NIV).

Having their foundation "outside the camp," the believers can renounce an enduring city in this world, hoping for another city "that is to come." That is a specific way to cope with Pascal's "vast middle" and its uncertainty, by making this uncertainty the real condition of faith, connected to the hope of a future space, as a new city. Therefore the believers can confess that, joining Jesus outside the camp, they are here and now "aliens and strangers on earth" (Heb. 11:13 NIV), crossing this earth like pilgrims. That does not mean a negative relationship to the present space in which they live. It is rather a distance (or with Ricoeur, a distanciation) allowing a positive attitude against this passing world, for example the very open practice of welcoming strangers and sharing with those suffering: "Keep on loving each other as brothers. Do not forget to welcome strangers, for by so doing some people have welcomed angels without knowing it. Remember those in prison as if you were their fellow-prisoners, and those who are ill-treated as if you yourselves were suffering" (Heb. 13:1–3 NIV).

Conclusion

I hope these reflections have shown that this book is full of hermeneutical challenges, both for a philosophical and a theological hermeneutics. As I already said in the introduction, the contributions, by their multiple interactions, create an interesting "space of interpretation" about the meaninglessness and meaningfulness of space! I tried to focus on some of the challenges, without pretending to be exhaustive. The rereading may go on, continued by other "rereaders" who will see other challenges. That also belongs to the hermeneutical condition of an open process of reception, provoked by a very enriching book!

Notes

1. Paul Ricoeur, *The Conflict of Interpretations: Essays in Hermeneutics*, Don Idhe, ed. (Evanston, IL: Northwestern University Press, 1974).
2. Maurice Merleau-Ponty, *Phenomenology of Perception* (New York: Routledge, 2012), First part, Chap. III; Second part, Chap. II.
3. Author's translation. Blaise Pascal, *Œuvres complètes*, Pref. Henri Gouhier, Intro., and Notes Louis Lafuma (Paris: Éditions du Seuil, 1963), 525–28. In Lafuma's numbering, it is thought 199, entitled "Disproportion of mankind" (in the more usual numbering of Brunschvicg: thought 434).
4. Pascal, *Œuvres complètes*, 527.
5. Pascal, *Œuvres complètes*, 527.
6. Merleau-Ponty, *Phenomenology of Perception*, Second part, Chap. I and Chap. II, D.
7. Paul Ricoeur, "The Hermeneutical Function of Distanciation," *Philosophy Today* 17.2 (1973): 129–41. Also to be found in Paul Ricoeur, *Hermeneutics and the Human Sciences: Essays on Language, Action, and Interpretation*, John B. Thompson, ed., trans., and intro. (Cambridge: Cambridge University Press, 1981).
8. Pascal, *Œuvres complètes*, 527.
9. Cf. Gerhard Ebeling, *Luther: An Introduction to His Thought* (London: Collins, 1970), mainly Chap. 12; Gerhard Ebeling, *Dogmatik des christlichen Glaubens*, 4th rev. ed.; 3 vols.; (Tübingen: Mohr Siebeck, 2012), 1.4 §14: "Der Mensch *coram Deo*," 334–55.

Bibliography

Abou-El-Haj, Barbara. "Artistic Integration and Social Consensus." In *Artistic Integration in Gothic Buildings*, edited by Virginia Chieffo Raguin, Kathryn Brush, and Peter Draper, 214–35. Toronto: University of Toronto Press, 1995.

Adams, Douglas. *The Hitchhiker's Guide to the Galaxy*. London: Pan Books, 1979.

Adams, Karin. "Metaphor and Dissonance: A Reinterpretation of Hosea 4:13–14." *Journal of Biblical Literature* 127 (2008): 291–305.

Aitken, Ellen Bradshaw. "The Landscape of Promise in the Apocalypse of Paul." In *Walk in the Ways of Wisdom: Essays in Honor of Elisabeth Schüssler FiorenzaI*, edited by Shelly Matthews, Cynthia Briggs Kittredge, and Melanie Johnston-DeBaufre, 153–65. Harrisburg, PA: Continuum International, 2003.

Akbar, Jamel. "Khat[.]t[.]a and the Territorial Structure of Early Muslim Towns." *Muqarnas* 6 (1989): 22–32.

Andersen, Francis and David N. Freedman. *Hosea: A New Translation with Introduction and Commentary*. Anchor Bible 24. Garden City, NY: Doubleday, 1980.

Antal, Frederick. *Hogarth and His Place in European Art*. London: Routledge & Kegan Paul, 1962.

Arnold, Edwin. *East and West, Being Papers Reprinted from the "Daily Telegraph" and Other Sources*. London: Longmans, Green and Company, 1896.

Appadurai, Arjun. *The Social Life of Things: Commodities in Cultural.* Perspective. New York: Cambridge University Press, 1986.

Arthur, Linda. *Religion, Dress and the Body*. Oxford: Berg, 1999.

———. *Undressing Religion: Commitment and Conversion from a Cross-Cultural Perspective*. Oxford: Berg, 2000.

Arukihenro—Walking Pilgrims, directed by Tommi Mendel. Ethnographic Documentary Film, 2006. Watertown, MA: Tigertoda Productions.

Assante, Julia. "From Whores to Hierodules: The Historiographic Invention of Mesopotamian Female Sex Professionals." In *Ancient Art and Its Historiography*, edited by Alice A. Donohue and Mark D. Fullerton, 13–47. Cambridge: Cambridge University Press, 2003.

Atherton, Herbert M. *Political Prints in the Age of Hogarth: A Study of the Ideographic Representation of Politics.* Oxford: Clarendon Press, 1974.

Auffarth, Christoph. *Irdische Wege und himmlischer Lohn: Kreuzzug, Jerusalem und Fegefeuer in religionswissenscahftlicher Perspektive.* Göttingen: Vandenhoeck and Ruprecht, 2002.

Baatz, Ursula, Hans Belting, und Isolde Charim, eds. *Bilderstreit 2006: Pressefreiheit? Blasphemie? Globale Politik?* Wien: Picus, 2007.

Badhra, Gautam, Gyan Prakash, and Susie Tharu, eds. *Subaltern Studies X.* New Delhi: Oxford University Press, 1999.

Barasch, Moshe. "Apocalyptic Space." In *Apocalyptic Time*, edited by Albert I. Baumgarten, 305–26. Leiden: Brill, 2000.

Barringer, Tim and Tom Flynn, eds. *Colonialism and the Object: Empire, Material Culture and the Museum.* London: Routledge, 2004.

Barnard, Malcolm. *Fashion as Communication.* New York: Routledge, 2008.

Bartetzko, Dieter. "Alarmstufe Kitsch." *Frankfurter Allgemeine Zeitung*, November 29, 2005.

Bashō, Matsuo. *The Narrow Road to the Deep North and Other Travel Sketches.* London: Penguin, 1966.

Baumann, Gerlinde. *Love and Violence: Marriage as Metaphor for the Relationship between YHWH and Israel in the Prophetic Books*, translated by Linda M. Maloney. Collegeville, MN: Liturgical Press, 2003.

Bautch, Kelley Coblentz. *A Study of the Geography of 1 Enoch 17–19: "No One Has Seen What I Have Seen."* Leiden: Brill, 2003.

Becker, Christoph. *Paul Gaugin—Tahiti.* Stuttgart: Verlag Gerd Hatje, 1998.

Beinhauer-Köhler, Bärbel, Daria Pezzoli-Olgiati, and Joachim Valentin, eds. *ReligiöseBlicke—Blicke auf das Religiöse. Visualität und Religion.* Zürich: TheologischerVerlag, 2010.

Belting, Hans. "Der Ort der Bilder." In *Das Erbe der Bilder: Kunst und modern Medien in den Kulturen der Welt*, edited by Hans Belting and Lydia Haustein, 34–53. München: C. H. Beck 1998.

———. *Florenz und Bagdad: Eine Westöstliche Geschichte des Blicks.* München: Beck, 2008.

Bennet, Tony. *Pasts Beyond Memory: Evolution, Museums, Colonialism.* London: Routledge, 2004.

Benz, Arnold. *Das geschenkte Universum—Astrophysik und Schöpfung*, 2nd edition. Düsseldorf: Patmos, 2010.

Berdoulay, Vincent. "Place, Meaning, and Discourse in French Language Geography." In *The Power of Place*, edited by John A. Agnew and James S. Duncan, 124–39. London: Unwin Hyman, 1989.

Berger, Hans. *Evangelisch Chur—Seine Prädikanten, Kirchen und Friedhöfe.* Chur: Calven, 1978.

Berminham, Ann. *Landscape and Ideology: The English Rustic Tradition, 1740–1860.* Berkeley: University of California Press, 1996.

Bernstein, Alan E. *The Formation of Hell: Death and Retribution in the Ancient and Early Christian Worlds*. Ithaca, NY: Cornell University Press, 1993.

Berquist, Jon L. and James W. Flanagan. "Ancient Perceptions of Space/ Perceptions of Ancient Space." *Semeia* 87 (1999): 15–43.

Berry, Lloyd E., ed. *The Geneva Bible: A Facsimile of the 1560 Edition*. Madison: University of Wisconsin Press, 1969.

Bertram, Jutta. "Frohe Feste voll überschäumender Lebensfreude. Die Rolle der Exotik im Dritte-Welt-Tourismus." In *Trouble in Paradise: Tourismus in die Dritte Welt*, edited by Christian Stock, 100–12. Düsseldorf: Haus Altenberg, 1997.

Binder, Jana. *Globality. Eine Ethnographie über Backpacker*. Münster: Lit, 2005.

Binski, Paul. *Medieval Death: Ritual and Representation*. Ithaca, NY: Cornell University Press, 1996.

Bird, Phyllis A. "'To Play the Harlot': An Inquiry into an Old Testament Metaphor." In *Missing Persons and Mistaken Identities: Women and Gender in Ancient Israel*, 219–36. Minneapolis: Fortress Press, 1997.

Blair, Sheila. *Islamic Inscriptions*. New York: New York University Press, 1998.

Blatti, J. O., ed. *Past Meets Present: Essays about Historic Interpretation and Public Audiences*. Washington, DC: Smithsonian Institution Press, 1987.

Blumenberg, Hans. *Die Lesbarkeit der Welt*. Frankfurt a.M.: Suhrkamp, 1981.

Bœspflug, François. *Caricaturer Dieu? Pouvoirs et dangers de l'image*. Paris: Bayard, 2006.

Böhme, Gernot. "Der Raumleiblicher Anwesenheit und der Raumals Medium der Darstellung." In *Performativität und Medialität*, edited by Sybille Krämer, 129–40. München: Wilhelm Fink, 2004.

Bordwell, David. *Narration in Fiction Film*. Madison: University of Wisconsin Press, 1985.

Bordwell, David and Kristin Thompson. *Film Art: An Introduction*. 8th edition. New York: McGraw-Hill, 2008.

Branigan, Edward. *Narrative Comprehension and Film*. London: Routledge, Taylor and Francis, 1992.

Bratton, Susan. *Environmental Values in Christian Art*. Albany: State University of New York Press, 2008.

Bruno, Giordana. *Atlas of Emotion: Journeys in Art, Architecture, and Film*. New York: Verso, 2002.

Buber, Martin. "The Spirit of the Orient and Judaism." *On Judaism*. 56–78. New York: Schocken, 1996.

Buck, Anne. *Victorian Costume and Costume Accessories*. Bedford: Carlton, 1984.

Bugslag, James. "Ideology and Iconography in Chartres Cathedral: Jean Clement and the Oriflamme." *Zeitschrift für Kunstgeschichte* 61.4 (1998): 491–508.

Bugslag, James. "Pilgrimage to Chartres: The Visual Evidence." In *Art and Architecture of Late Medieval Pilgrimage in Northern Europe and the British Isles*, edited by Sarah Blick and Rita Tekippe, 135–83. Leiden: Brill, 2005.

Burckhardt, Titus and Roland Michaud. *Art of Islam: Language and Meaning*. London: World of Islam Festival Pub. Co, 1976.

Bynum, Caroline Walker. *The Resurrection of the Body in Western Christianity, 200–1336*. New York: Columbia University Press, 1995.

Byrne, Denis. "Buddhist Stupa and Thai Social Practice." *World Archaeology* 27.2 (1985): 266–81.

Camille, Michael. *Image on the Edge: The Margins of Medieval Art*. Cambridge, MA: Harvard University Press, 1992.

Carrol, Noël. *The Philosophy of Motion Pictures*. Malden, MA: Blackwell, 2008.

Casey, E. S. *The Fate of Place*. Berkeley: University of California Press, 1997.

Cassirer, Ernst. *An Essay on Man*. New Haven, CT: Yale University Press, 1944.

Castells, Manuel. *The Informational City*. Oxford: Basil Blackwell, 1989.

Cetina, Karin Knorr. "'Viskurse' der Physik: Wie visuelle Darstellungen ein Wissenschaftsgebiet ordnen." In *Konstruktionen Sichtbarkeiten*, edited by Jörg Huber and Martin Heller, 245–63. New York: Springer, 1999.

Chakrabarty, Dipesh. *Habitations of Modernity: Essays in the Wake of Subaltern Studies*. Chicago: University of Chicago Press, 2002.

Chaturvedi, Vinayak, ed. *Mapping Subaltern Studies and the Postcolonial*. London: Verso, 2000.

Chédeville, André. *Chartres et ses Compagnes (XIe–XIIIe siècles)*. Paris: C. Klincksieck, 1973.

Cherniss, Harold. *Aristotle's Criticism of Plato and the Academy*. New York: Praeger, 1962.

Choay, Françoise. *L'urbanisme, utopies et réalités, Une anthologie*. Paris: Editions du Seuil, 1965.

Clifford, James. "On Collecting Art and Culture." In *The Predicament of Culture: Twentieth Century Ethnography, Literature, and Art*, edited by James Clifford, 215–51. Cambridge, MA: Harvard University Press, 1988.

Cohen, Erik. "Pilgrimage and Tourism: Convergence and Divergence." In *Sacred Journeys: The Anthropology of Pilgrimage*, edited by Alan Morinis, 47–65. Westport, CT: Greenwood Press, 1992.

Cohn, Bernard S. *Colonialism and Its Forms of Knowledge: The British in India*. Princeton, NJ: Princeton University Press, 1996.

Coleman, Simon and John Elsner. *Pilgrimage: Past and Present in the World Religions*. Cambridge, MA: Harvard University Press, 1995.

Coleman Simon and John Eade, eds. *Reframing Pilgrimage: Cultures in Motion*. London: Routledge, 2004.

Collins, John J. "Introduction: Towards the Morphology of a Genre." *Semeia* 14 (1979): 1–20.

Collon, Dominique, Sara Peterson, Eva Wilson, John Villers, Eva Baer, and W. Iain MacKay. "Ornament and Pattern." In *Grove Art Online*. Oxford University Press, 2007–. Accessed June 28, 2013, http://www.oxfordartonline.com/subscriber/article/grove/art/T063880.

Copeland, Kirsti. "Thinking with Oceans: Muthos, Revelation, and the Apocalypse of Paul." In *The Visio Pauli and the Gnostic Apocalypse of Paul*, edited by Jan N. Bremmer and István Czachesz, 77–104. Leuven: Peeters, 2007.

Corry, John A. *Satirical View of London: Comprehending a Sketch of the Manners of the Age*. London: J. Morton, 1803.

Craik, Jennifer. *The Face of Fashion: Cultural Studies in Fashion*. New York: Routledge, 1994.

Dabydeen, David. *Hogarth's Blacks: Images of Blacks in Eighteenth Century English Art*. Mundelstrup: Dangaroo Press, 1985.

Dahinden, Janine, Anna-Katharina Höpflinger and Anne Lavanchy, eds. "Gender and Religion: Towards Diversity." *Special Issue of Women's Studies: An Interdisciplinary Journal* 41.6 (2012).

Dalché, Patrick Gautier. *La "Descriptio Mappae Mundi" de Hugues de Saint-Victor: Texte inédite avec introduction et commentaire*. Paris: Études Augustiniennes, 1988.

Davis, Fred. "Clothing and Fashion as Communication." In *The Psychology of Fashion*, edited by Michael R. Solomon, 15–27. Lexington, MA: Lexington Books, 1985.

Davis, Richard. *The Lives of Indian Images*. Princeton, NJ: Princeton University Press, 1997.

Day, John. "Asherah." In Freedman, *Anchor Bible Dictionary*, 1:486.

Debatin, Bernhard, ed. *Der Karikaturenstreit und die Pressefreiheit, Wert- und Normenkonflikte in der globalen Medienkultur*. Berlin: Lit Verlag, 2007.

De Certeau, Michel. *The Practice of Everyday Life*. Berkeley: University of California Press, 1984.

Deeken, Annette. *Reisefilme: Ästhetik und Geschichte*. Filmstudien 38. Remscheid: Gardez!, 2004.

De los Ríos, José Amador and Pedro de Madrazo. *El estilo Mudejar en arquitectura: discurso*. Valencia: Librerías "Paris-Valencia," 1996.

De Lubac, Henri. *Corpus Mysticum: L'Eucharistie et l'Église au Moyen Age*. Paris: Aubier-Montaigne, 1949.

de Offray, Julien. *L'homme machine*. Charleston, SC: Nabu Press, 2010.

Die Frühzeit des Holzschnitts. München: Staatliche Graphische Sammlung München, 1970. Exhibition catalog.

Dirks, Nicholas B. *Castes of Mind: Colonialism and the Making of Modern India*. Princeton, NJ: Princeton University Press, 2001.

Dodds, Jerrilynn Denise. *Al-Andalus: The Art of Islamic Spain*. New York: Metropolitan Museum of Art: Distributed by H. N. Abrams, 1992.

Dodds, Jerrlynn Denise, Maria Rosa Menocal, and Abigail Krasner Balbale. *The Arts of Intimacy: Christians, Jews, and Muslims in the Making of Castilian Culture*. New Haven, CT: Yale University Press, 2008.

Donald, Diana. *The Age of Caricature: Satirical Prints in the Reign of George III.* New Haven, CT: Yale University Press, 1996.

Doyle, Tara. "Bodh Gayā: Journeys to the Diamond Throne and the Feet of Gayāsura." PhD diss., Harvard University, 1997.

Dubisch, Jill and Michael Winkelmann, eds. *Pilgrimage and Healing.* Tucson: University of Arizona Press, 2005.

Dunning, William. *Changing Images of Pictorial Space: A History of Spatial Illusion in Painting.* Syracuse: Syracuse University Press, 1991.

Durham, John I. *Exodus.* Word Biblical Commentary. Waco, TX: Word, 1987.

Eco, Umberto. *The Infinity of Lists.* New York: Rizzoli, 2009.

Eliade, Mircea. *Das Heilige und das Profane: Vom Wesen des Religiosen.* Hamburg: Rowohlt, 1957.

Erickson, Thomas. "Between Universalism and Relativism: A Critique of the UNESCO Concept of Culture." *Culture and Rights: Anthropological Perspectives,* edited by Jane Cowan, 127–48. Cambridge: Cambridge University Press, 2001.

Errington, Shelly. "Making Progress on Borobudur: An Old Monument in New Order." *Visual Anthropology Review* 9.2 (1993): 32–59.

Falchetta, Piero. *Fra Mauro's World Map.* Turnhout: Brepols, 2006.

Farinelli, Franco. *Geografia: Un'introduzione ai modelli del mondo.* Torino: Einaudi, 2003.

———. *I segni del mondo: Immagine cartografica e discorso geografico in età moderna.* Firenze: La Nuova Italia, 1992.

———. *L'invenzione della Terra.* Palermo: Sellerio, 2007.

Fassler, Margot. "Liturgy and Sacred History in the Twelfth-Century Tympana at Chartres." *The Art Bulletin* 75.3 (1993): 499–520.

Favret-Saada, Jeanne. *Comment produire une crise mondiale avec douze petits dessins.* Paris: Les Prairies ordinaires, 2007.

Fielding, Henry. *The History of the Adventures of Joseph Andrews,* edited by Adam Potkay. New York: Pearson/Longman, 2010.

Fluck, Winfried. "Imaginary Space; or, Space as Aesthetic Object." In *Space—Place—Environment,* edited by Lother Hönnighausen, Julia Apitzsch, Wibke Reger, 25–40. Tübingen: Stauffenburg Verlag, 2004.

Frassetto, Michael. "Resurrection of the Body: Eleventh-Century Evidence from the Sermons of Ademar of Chabannes." *Journal of Religious History* 26.3 (2002): 235–49.

Frazer, Sir James. *The Golden Bough: A Study in Comparative Religion.* 2 vols. London: Macmillan, 1890.

Freedman, David Noel. *The Anchor Bible Dictionary.* 6 vols. New York: Doubleday, 1992.

Fretheim, Terence E. *Exodus.* Interpretation: A Bible Commentary for Teaching and Preaching. Louisville, KY: John Knox Press, 1991.

Frey, Nancy Louise. Pilgrim Stories. *On and off the Road to Santiago.* Berkeley: University of California Press, 1998.

Frohnmeyer, L. J. and I. Benzinger. *Bilderatlas zur Bibelkunde: Ein Handbuch für den Religionslehrer und Bibelfreund*. Stuttgart: Theodor Benzinger, 1905.

Gatta, John. *Making Nature Sacred*. Oxford: Oxford University Press, 2004.

Garland, Alex. *The Beach*. London: Penguin, 2007.

Gebser, Jean. *The Ever-Present Origin*. Athens: Ohio University Press, 1985.

Geertz, Clifford. "Religion as a Cultural System." In *The Interpretation of Cultures: Selected Essays*, 87–125. New York: Basic Books, 1973.

General Conference of the United Nations Educational, Scientific and Cultural Organisation. "Convention Concerning the Protection of the World Cultural and Natural Heritiage." Paris, November 16, 1972. Accessed July 1, 2013, http://whc.unesco.org/en/conventiontext/.

———. "Convention for the Safeguarding of Intangible Cultural Heritage." Paris, October 2003. Accessed July 1, 2013, http://www.unesco.org /culture/ich/en/convention.

.——— "Information Dossier for nomination of Mahabodhi Temple Complex, Bodhgaya as a World Heritage Site." Accessed July 1, 2013, http://whc.unesco.org/uploads/nominations/1056rev.pdf.

George, Mark. *Israel's Tabernacle as Social Space*. Society of Biblical Literature Ancient Israel and Its Literature 2. Atlanta, GA: Society of Biblical Literature, 2009.

Glavac, Monkia. *Der Fremde in der europäischen Karikatur, Eine Studie über das Spannungsfeld zwischen Belustigung, Beleidigung und Kritik*. Göttingen: Vandenhoeck und Ruprecht, 2012.

Gleis, Matthias. *Die Bamah*. Beiheft zur Zeitschrift für die alttestamentliche Wissenschaft 251. Berlin: De Gruyter, 1997.

Goldstein, Justin Robert. *Censorship of Political Caricature in Nineteenth-Century France*. Kent, OH: Kent State University Press, 1989.

———. *Political Censorship of the Arts and the Press in Nineteenth-Century Europe*. New York: St. Martin's Press, 1989.

Gombrich, E. H. *The Sense of Order: A Study in the Psychology of Decorative Art*, 2nd edition. London: Phaidon Press, 1984.

Gorman Jr., Frank H. *The Ideology of Ritual: Space, Time and Status in the Priestly Theology*. Journal for the Study of the Old Testament: Supplement Series 91. Sheffield: JSOT Press, 1990.

Grabar, Oleg. *The Mediation of Ornament*. Princeton, NJ: Princeton University Press, 1992.

Guardini, Romano. *Vom Geist der Liturgie*. Freiburg i.B: Herdersche Verlagshandlung, 1918.

Gunning, Tom. "Traveling Shots." In *Traveling Shots: Film als Kaleidoskop von Reiseerfahrungen*, edited by Winfried Pauleit, Christine Rüffert, Karl-Heinz Schmid, Alfred Tews, and Bremer Symposium zum Film, 16–29. Berlin: Bertz und Fischer, 2007.

Guth, Alan H. *The Inflationary Universe: The Quest for a New Theory of Cosmic Origins*. Boston, MA: Addison-Wesley, 1997.

Haab, Barbara. *Weg und Wandlung*. Freiburg: Universitätsverlag Freiburg Schweiz, 1998.

Hahn-Woernle, Birgit. *Die Erbstorfer Weltkarte*. Ebstorf: Kloster, 1987.

Hall, Stuart. *Representation: Cultural Representations and Signifying Practices*. London: Sage, 1997.

Halm, Philipp M. "Ikonographische Studien zum Armen Seelen-Kultus. Mit 15 Abbildungen im Text." *Münchener Jahrbuch der bildenden Kunst* 12 (1921–22): 1–24.

Hannoosh, Michele. "Caricature." *Encylopedia of Aesthetics: Oxford Art Online*. Oxford University Press. Accessed September 7, 2010, http://www.oxfordartonline.com/subscriber/article/opr/t234/e0102.

Harvey, L. P. *Islamic Spain, 1250 to 1500*. Chicago: University of Chicago Press, 1990.

Hassig, Debra. "The Iconography of Rejection: Jews and Other Monstrous Races." In *Image and Belief: Studies in Celebration of the Eightieth Anniversary of the Index of Christian Art*, edited by Colum Hourihane, 25–46. Princeton, NJ: Princeton University Press, 1999.

Heidegger, Martin. "Bauen, Wohnen, Denken." In *Vorträge und Aufsätze*, 145–62. Pfullingen: Neske, 1954.

Hennig, Christoph. *Reiselust, Toursimus, Tourismus und Urlaub*. Frankfurt a.M.: Suhrkamp, 1999.

Hereafter, directed by Clint Eastwood. 2010. Burbank, CA: Warner Brothers.

Hervieu-Léger, Danièle. "Space and Religion: New Approaches to Religious Spatiality in Modernity." *International Journal of Urban and Regional Research* 26.1 (2002): 99–105.

Hickethier, Knut. "Zur Analyse des Visuellen, des Auditiven und des Narrativen." In *Film- und Fernsehanalyse*, 3rd edition, 39–89. Stuttgart: J. B. Metzler, 2001.

Himmelfarb, Martha. *Ascent to Heaven in Jewish and Christian Apocalypses*. New York: Oxford University Press, 1993.

Hollengreen, Laura. "From Medieval Sacred Place to Modern Secular Space: Changing Perspectives on the Cathedral and Town of Chartres." In *Architecture as Experience: Radical Change in Spatial Practice*, edited by Dana Arnold and Andrew Ballantyne, 81–108. New York: Routledge, 2004.

Höpflinger, Anna-Katharina. "From Visibility to Invisibility. A Semiotic Approach to Marduk's Battle against Tiamat on a Relief from Palmyra." In *Approaches to the Visual in Religion*, edited by Daria Pezzoli-Olgiati and Christopher Rowland, 39–50. Göttingen: Vandenhoeck and Ruprecht, 2011.

———."'Mehr verschandelt als verwandelt'. Kleidung als Medium der Geschlechterkonstruktion in religiösen Symbolsystemen." In *Handbuch*

Gender und Religion, edited by Anna-K. Höpflinger, Ann Jeffers, and Daria Pezzoli-Olgiati, 243–55. Göttingen: Vandenhoeck und Ruprecht, 2008.

———. *Schlangenkampf: Ein Vergleich von ausgewählten Bild- und Textquellen aus dem griechisch-römischen und dem altorientalischen Kulturraum*. Zürich: TVZ, 2010.

Hoskin, Michael. *The Cambridge Concise History of Astronomy*. Cambridge: Cambridge University Press, 1999.

Hugh of Saint-Victor. *Opera Omnia*, II. Paris: Migne, 1880.

Hurowitz, Victor Avigdor. *I Have Built You an Exalted House: Temple Building in the Bible in Light of Mesopotamian and Northwest Semitic Writings*. Journal for the Study of the Old Testament: Supplement Series 115. Sheffield: JSOT Press, 1992.

Hüttner, J. C. "Die jetzigen Karikaturzeichner in London." In *London und Paris, 1798–1815, hrsg. von F. J. Bertuch, Bd. 18* (1806). Rudolstadt: 7–10. Accessed April 9, 2010, http://www.ub.uni-bielefeld.de/diglib/aufkl/londonparis/index.htm.

Huxley, Julian. *UNESCO: Its Purpose and Philosophy*. Washington, DC: American Council on Public Affairs, 1947.

International Council on Monuments and Sites. "Mahabodhi Temple Complex at Bodh Gaya (India) 1056." Accessed July 1, 2013, http://whc.unesco.org/archive/2002/whc-02-conf201-inf2adde.pdf.

Jarrassé, Dominique. *Synagogues: Architecture and Jewish Identity*. Paris: Vilo, 2001.

Jezler, Peter, ed. *Hillem, Hölle, Fegefeuer: Das Jenseitsim Mittelalter*. Zürich: Neue Zürcher Zeitung, 1994.

Joos, Raimund and Michael Kasper. *Outdoor: Der Weg ist das Ziel. Spanien: Jakobsweg. Camino Francés*. Welver: Conrad Stein, 2009.

Jung, C.G. and Richard Wilhelm, eds. *Geheimnis der Goldenen Blüte*. München: Diederichs, 1990.

Kantorowicz, Ernst. *The King's Two Bodies: A Study in Medieval Political Theology*. Princeton, NJ: Princeton University Press, 1957.

Kemp, Wolfgang. "Kunstwissenschaft und Rezeptionsästhetik." In *Der Betrachter ist im Bild: Kunstwissenschaft und Rezeptionsästhetik*, edited by Wolfgang Kemp, 7–29. Berlin: Dietrich Reimer Verlag, 1992.

Kennedy, Archibald R. S. "Tabernacle." In *A Dictionary of the Bible Dealing with Its Language, Literature and Contents Including the Biblical Theology*, vol. IV, edited by James Hastings. New York: Charles Scribner's Sons, 1902.

Kerkeling, Hape. *Ich bin dann mal weg: Meine Reise auf dem Jakobsweg*. München: Piper, 2006.

Kerouac, Jack. *On the Road: The Original Scroll*. New York: Penguin, 2007.

Kessler, Herbert L. *Seeing Medieval Art*. Toronto: University of Toronto Press, 2004.

Kibler, William W. and Grover Zinn. *Medieval France*. New York: Garland, 1995.

Klausen, Jytte. *The Cartoons That Shook the World.* New Haven, CT: Yale University Press, 2009.

Kohl, Karl-Heinz. "Die Magie des Okzidents: Europa in den Berichten arabischer und chinesischer Reisender." In *Der Orient im Okzident, Sichtweisen und Beeinflussungen*, edited by Irene A. Diekmann, Thomas Gerber, Julius H. Schoeps, 77–93. Potsdam: Verlag für Berlin-Brandenburg, 2003.

Krysmanski, Bernd W. *Hogarth's Enthusiasm Delineated: Nachahmung als Kritik am Kennertum, Eine Werkanalyse, 2 Binding.* New York: Georg Olms Verlag, 1996.

Kunelius, Risto and Elisabeth Eide. "The Mohammed Cartoons, Journalism, Free Speech and Globalization." In *Reading the Mohammed Cartoons Controversy: An International Analysis of Press Discourses on Free Speech and Political Spin*, edited by Risto Kunelius, 9–21. Bochum/Freiburg: Projekt Verlag, 2007.

Kunz, Ralph and Thomas Schlag. "Des Pfarrers neue Kleider . . . Erkundungen und Erwägungen zur Talarfrage in den reformierten Landeskirchen." *Pastoraltheologie* 98.6 (2009): 212–32.

Kurmann, Peter and Brigitte Kurmann-Schwartz. "Chartres Cathedral as a Work of Artistic Integration: Methodological Reflections." In *Artistic Integration in Gothic Buildings*, edited by Virginia Chieffo Raguin, Kathryn Brush and Peter Draper, 131–52. University of Toronto Press, 1995.

Lacoste, Jean-Yves. *Expérience et absolu: Questions disputées sur l'humanité de l'homme.* Paris: Presses Universitaires de France, 1994.

Lanwerd, Susanne. *Religionsästhetik: Studienzum Verhältnis von Symbol und Sinnlichkeit.* Würzburg: Königshausen and Neumann, 2002.

Largo, R. H. *Babyjahre.* München: Piper, 2007.

Largo, R. H. and J. A. Howard. "Developmental Progression of Play Behaviors in Children 9 to 30 Months of Age. I. Spontaneous Play and Imitation." *Developmental Medicine and Child Neurology* 21 (1979a): 299–310.

———. "Developmental Progression of Play Behaviors in Children 9 to 30 Months of Age. ii. Spontaneous Play and Language Development." *Developmental Medicine and Child Neurology* 21 (1979b): 492–503.

Lefebvre, Henri. *The Production of Space,* translated by Donald Nicholson-Smith. Malden, MA: Blackwell, 1991.

Le Goff, Jacques. *The Medieval Imagination,* translated by Arthur Goldhammer. Chicago: University of Chicago Press, 1992.

Lermack, Annette. "Spiritual Pilgrimage in the Psalter of Bonne of Luxembourg." In *The Art, Science, and Technology of Medieval Travel*, edited by Robert Bork and Andrea Kann, 97–111. Burlington, VT: Ashgate, 2008.

Levenson, Jon D. *Creation and the Persistence of Evil: The Jewish Drama of Divine Omnipotence.* San Francisco: Harper & Row, 1988.

Levinas, Emmanuel. *Totalité et infini*. Nijhoff: La Haye, 1971.

Lévi-Strauss, Claude. *Traurige Tropen*. Köln: Suhrkamp, 1978.

Lied, Liv Ingeborg. "Another Look at the Land of Damascus: The Spaces of the Damascus Document in the Light of Edward W. Soja's Thirdspace Approach." In *New Directions in Qumran Studies: Proceedings of the Bristol Colloquium on the Dead Sea Scrolls, 8–10 September 2003*, edited by Jonathan Goodson Campbell, William John Lyons, and Lloyd Pietersen, 101–25. London: Continuum International Publishing Group, 2005.

Lilley, Keith D. "Cities of God? Medieval Urban Forms and Their Christian Symbolism." *Transactions of the Institute of British Geographers* 29.3 (2004): 296–313.

Lloyd, Seth. *Programming the Universe: A Quantum Computer Scientist Takes On the Cosmos*. New York: Alfred A. Knopf, 2006.

Lopez, Kathryn M. "Standing before the Throne of God: Critical Spatiality in Apocalyptic Scenes of Judgment." In *Constructions of Space II: The Biblical City and Other Imagined Spaces*, edited by Jon L. Berquist and Claudia V. Camp, 139–55. New York: T&T Clark, 2008.

Löw, Martina. "Die topologischen Dimensionender Kultur." In *Handbuch der Kulturwissenschaften: vol. 1: Grundlagen und Schlüsselbegriffe*, edited by Friedrich Jaeger and Burkhard Liebsch, 46–59. Stuttgart: Mezler, 2004.

Ludden, David, ed. *Reading Subaltern Studies: Critical History, Contested Meaning and the Globalization of South Asia*. London: Anthem South Asian Studies 2002.

Lurie, Alison. *The Language of Clothes*. London: Bloomsbury, 1992.

Lyden, John C. *Film as Religion: Myths, Moral, and Rituals*. London: New York University Press, 2003.

Mäder, Marie-Therese. "A Cultural Studies Approach to Film and Religion, Context and Film Analysis of Yes" (Potter, GB/USA 2004). In *Approaches to the Visual in Religion*, edited by Christopher Rowland and Daria Pezzoli-Olgiati, 101–18. Göttingen: Vandenhoeck and Ruprecht, 2011.

———. *Die Reise als Suche nach Orientierung, Eine Annäherung and das Verhältnis zwischen Religion und Film*. Marburg: Schüren 2012.

Magnaghi, Alerbto. *Il progetto locale: Verso la coscienza di luogo*. Torino: Bollati Boringhieri, 2001.

Mahabodhi. "The Bodhgaya Temple Act, 1949." Accessed July 1, 2013, http://www.mahabodhi.com/en/b-gaya_tempact.htm.

Majid, Anouar. *We Are All Moors: Ending Centuries of Crusades against Muslims and Other Minorities*. Minneapolis: University of Minnesota Press, 2009.

Mäkipä, Tea. "World of Plenty." Digital Print 400 × 2670 cm, 2005. World Expo exhibition in Aichi, Tea.

Mâle, Emile. *The Gothic Image: Religious Art in France of the Thirteenth Century*. Boulder, CO: Westview, 1972.

Malik, Jamal, Jörg Rüpke and Theresa Wobbe, eds. *Religion und Medien: Vom Kultbild zum Internet*. Münster: Aschendorff, 2007.

Malpas, Jeff. *Place and Experience*. Cambridge: Cambridge University Press, 1999.

Már Jónsson, Einar. *Le miroir: Naissance d'un genre littéraire*. Paris: Les belles lettres, 1995.

Marc'hadour, Germain and Roland Galicois, eds. *Erasme de Rotterdam et Thomas More: Correspondance*. Sherbrooke: Université de Sherbrooke, 1985.

Marchetti, Paolo. *De Iure Finium: Diritto e confini tra tardo medioevo ed età moderna*. Milano: Giuffré, 2001.

Marin, Louis. *Utopiques: Jeux d'espaces*. Paris: Les Éditions de Minuit, 1973.

Martig, Charles and Daria Pezzoli-Olgiati, eds. *Outer Space, Reisen in Gegenwelten: Film und Theologie 12*. Marburg: Schüren, 2009.

Massey, Doreen, Pat Jess, and Jess Massey, eds. *A Place in the World? Places, Culture and Globalisation*. Oxford: Oxford University Press, 1995.

McCreery, Cindy. *The Satirical Gaze: Prints of Women in Late Eighteenth-Century England*. Oxford: Clarendon Press, 2004.

McNeil, Peter and Giorgio Riello. *Shoes: A History from Sandals to Sneakers*. New York: Berg, 2006.

———. "The Art and Science of Walking: Mobility, Gender and Footwear in the Long Eighteenth Century." *Fashion Theory 9* (2005): 175–204.

Melancholia, directed by Lars von Trier. 2011. Hvidover, Denmark: Zentropa.

Melchiorre, Virgilio. *Corpo e persona*. Milano: Vita e Pensiero, 1970.

———. *La coscienza utopica*. Milano: Vita e Pensiero, 1970.

Mendel, Tommi. "Arukihenro, Reflections on Fieldwork by a Visual Anthropologist." Electronic Journal of Contemporary Japanese Studies (2008): Discussion Paper No. 6, accessed October 30, 2011, http://www.japanesestudies.org.uk/discussionpapers/2008/Mendel.html.

———. "Foot-Pilgrims and Backpackers: Contemporary Ways of Travelling." In *Pilgrimage Today*, edited by Tore Ahlbäck and Björn Dahla, 288–315. Åbo: Donner Institute for Research in Religious and Cultural History, 2010.

———. "Kleidung und Körpererfahrung als Faktoren von Transitory Travelling Communities. Temporäre Identitäsbildung unter Fusspilgern in Japan und Spanien." In *Second Skin. Kleidung und Körper als Repräsentation von Religion*, edited by Monika Glavac, Anna-Katharina Höpflinger, and Daria Pezzoli-Olgiati, 53–78. Göttingen: Vandenhoeck & Ruprecht, 2013.

Menocal, Maria Rosa. *The Ornament of the World: How Muslims, Jews, and Christians Created a Culture of Tolerance in Medieval Spain*. Boston, MA: Little, Brown, 2002.

Merleau-Ponty, Maurice. *Phénoménologie de la perception*. Paris: Gallimard, 1945.

Meyer, Birgit, ed. *Aesthetic Formations. Media, Religion, and the Senses.* New York: Palgrave, 2009.

Meyers, Carol L. *Exodus.* New York: Cambridge University Press, 2005.

Milgrom, Jacob. *Leviticus 1–16: A New Translation with Introduction and Commentary.* Anchor Bible 3. New York: Doubleday, 1991.

Minois, Georges. *Historie des enfers.* Paris: Fayard, 1991.

Misitelli, Jean. "World Heritage: Between Universalism and Globalization." *International Journal of Cultural Property* 11.2 (2002): 323–36.

More, Thomas. *Complete Works*, IV. New Haven, CT: Yale University Press, 1965.

Morgan, David, ed. *Religion and Material Culture: The Matter of Belief.* New York: Routledge, 2010.

———. *The Sacred Gaze. Religious Visual Culture in Theory and Practice.* Los Angeles: University of California Press, 2005.

Muller, Gerhard, ed. "Paradise, Religionsgeschichtlich." In *Theologische Realenzyklopädie*, vol. 25, 705–8. Berlin: Walter De Gruyter Inc, 1995.

Münsterberg, Hugo. *Das Lichtspiel: Eine psychologische Studie (1916) und andere Schriften zum Kino*, edited and translated by Jörg Schweinitz. Wien: Synema, 1996.

Musitelli, Jean. "World Heritage: Between Universalism and Globalization." *International Journal of Cultural Property* 11.2 (2002): 323–36.

Nasr, Seyyed Hossein. "The Spiritual Message of Islamic Calligraphy." In *Religion, Art, and Visual Culture: A Cross-Cultural Reader*, edited by S. Brent Plate, 112–18. New York: Palgrave, 2002.

Novalis. *Die Christenheit oder Europa.* Stuttgart: Reclam, 1962.

Odermatt-Bürgi, Regula. "'SpeculumRationis' aus dem Beinhaus von Oberägeri." In *Unterwegs. Religion in Kunst und Brauchtum*, edited by Christine Kamm-Kyburz, Eugen Holz, and Richard Kern. Zug: Die Kunstgesellschaft, 1984.

Odin, Roger. "Kunst und Ästhetik bei Film und Fernsehen." In *Montage A/V, Zeitschrift für Theorie und Geschichte audiovisueller Kommunikation, 11/2/2002*, 42–57. Pragmatik des Films. Marburg: Stoke, 2002.

Olson, Vibeke. "Movement, Metaphor and Memory: The Interactions between Pilgrims and Portal Programs." In *Push Me, Pull You: Physical and Spatial Interaction in Late Medieval and Renaissance Art*, vol. 2, edited by Sarah Blick and Laura D. Gelfand, 495–521. Leiden: Brill, 2011.

O'Reilly, Camille Caprioglio. "From Drifter to Gap Year Tourist. Mainstreaming Backpacker Travel." *Annals of Tourism Research* 33 (2006): 998–1017.

Ostwalt, Conrad. "Teaching Religion and Film. A Fourth Approach." In *Teaching Religion and Film*, edited by Gregory J. Watkins, 35–54. Oxford: Oxford University Press, 2008.

Owens, Bruce McCoy. "Monumentality, Identity, and the State: Local Practice, World Heritage, and Heterotopia at Swayambhu, Nepa." *Anthropological Quarterly* 75.2 (Spring 2002): 269–316.

Pascal, Blaise. *Œuvres complètes.* Preface by Henri Gouhier. Introduction and notes by Louis Lafuma. Paris: Éditions du Seuil, 1963.

Pauleit, Winfried, Christine Rüffert, Karl-Heinz Schmid, Alfred Tews, and Bremer Symposium zum Film, eds. *Traveling Shots: Film als Kaleidoskop von Reiseerfahrungen.* Berlin: Bertz und Fischer, 2007.

Paulson, Ronald. *Hogarth, High Art and Low, 1732–1750,* vol. 2. Cambridge: Lutterworth, 1992.

———. *Hogarth's Graphic Works.* London: The Print Room, 1989.

———. *Hogarth, The Modern Moral Subject, 1697–1732,* vol. 1. Cambridge: Lutterworth, 1992.

Peirce, Charles Sanders. *Collected Papers,* edited by Charles Hartshorne and Paul Weiss. Cambridge, MA: The Belknap Press of Harvard University Press, 1966.

Peteu, Mihaela Cornelia and Sally Helvenston Gray. "Clothing Invention. Improving the Functionality of Women's Skirts, 1846–1920." *Clothing and Textiles Research Journal* 27 (2009): 45–61.

Pezzoli-Olgiati, Daria. "Approaching Afterlife Imagery: A Contemporary Glance at Ancient Concepts of Otherworldly Dimensions." In *Other Worlds and Their Relation to This World: Early Jewish and Ancient Christian Traditions,* edited by Tobias Niklas, 1–15. Leiden: Brill 2010.

———. "Eine illustrierte Annäherung an das Verhältnis von Medien und Religion." In *ReligiöseBlicke.* 245–66. Zurich: Theologischer Verlag Ag, 2010.

———. "Film und Religion: Blick auf Kommunikationssysteme und ihre vielfältigen Wechselwirkungen." In *Religious Turns—Turning Religions: Veränderte kulturelle Diskurse—Neue religiöse Wissensformen,* edited by Andreas Nehring and Joachim Valentin, 45–66. ReligionsKulturen 1. Stuttgart: Kohlhammer, 2008.

Pezzoli-Olgiati, Daria and Fritz Stolz. *Cartografia religiosa—Religiöse Kartographie—Cartographie religieuse.* Bern: Lang, 2000.

Pippen, Tina. "The Ideology of Apocalyptic Space." In *Constructions of Space II: The Biblical City and Other Imagined Spaces,* edited by Jon L. Berquist and Claudia V. Camp, 156–70. New York: T&T Clark, 2008.

Plate, S. Brent. "El mundo de Infante a través de la simbología universal de las estrellas." In *La Casa de Blas Infante en Coria del Río,* edited by Alberto Egea, 171–84. Sevilla: Centro de Estudios Andaluces, 2005.

———, ed. *Representing Religion in Worldcinema: Filmmaking, Mythmaking, Culture Making.* New York: Palgrave Macmillan, 2003.

Porter, Arthur Kingsley. *Medieval Architecture: Its Origins and Development, with Lists of Monuments and Bibliographies,* vol. 2. New York: Baker and Taylor, 1909.

Powell, Marvin A. "Weights and Measures." In Freedman, *Anchor Bible Dictionary,* 6: 897–908.

Propp, William H. C. *Exodus 19–40: A New Translation with Introduction and Commentary.* Anchor Bible 2A, edited by David Noel Freedman. New York: Doubleday, 2006.

Raguin, Virginia Chieffo and Sarah Stanbury. *Women's Space: Patronage, Place, and Gender in the Medieval Church*. Albany, NY: SUNY Press, 2005.

Rasche, Adelheid and Gundula Wolter. *Ridikül: Mode in der Karikatur 1600 bis 1900*. Berlin: SMB/Du Mont, 2003.

Rau, Cordula, ed. *Sehnsucht: The Book of Architectural Longings*. Wien: Springer Vienna Architecture, 2010.

Reader, Ian and Tony Walter, eds. *Pilgrimage in Popular Culture*. New York: Palgrave, 1993.

Reale, Giovanni. *Per una Nuova Interpretazione di Platone: Rilettura della Metafisica dei Grandi Dialoghi alla luce delle "Dottrine non Scritte."* Milano: Vita e Pensiero, 1997.

———. *Platone: Alla Ricerca della Sapienza Segreta*. Milano: Rizzoli, 1998.

Recht, Roland. *Believing and Seeing: The Art of Gothic Cathedrals*. Chicago: University of Chicago Press, 2008.

Reichen, Quirinus and Karen Christie, eds. *Das Schnittmusterbuch von Salomon Erb, "Livre des Chefs d'Oevre de la Maistrise des Tailleurs de Berne, 1730."* Bern: Bernisches Museum, 2000.

Richards, Greg and Julie Wilson, eds. *The Global Nomad. Backpacker Travel in Theory and Practice*. Clevedon, UK: Channel View Publications, 2009.

Ricoeur, Paul. *La métaphore vive*. Paris: Editions du Seuil, 1975.

Ritter, Carl. *Einleitung zur allgemeinen vergleichenden Geographie, und Abhandlungen zur Begründung einer mehr wissenschaftlichen Behandlung der Erdkunde*. Berlin: Reimer, 1852.

Roberson, Jennifer. "Visions of Al-Andalus in Twentieth-Century Spanish Mosque Architecture." In *Revisiting Al-Andalus: The Medieval and Early Modern Iberian World*, edited by Glaire D. Anderson and Mariam Rosser-Owen, 247–71. Leiden: Brill, 2007.

Rowland, Christopher. *The Open Heaven*. New York: Crossroad, 1982.

Rubinstein, Ruth P. *Dress Codes: Meanings and Messages in American Culture*. Boulder, CO: Westview Press, 1995.

Ruoff, Jeffrey, ed. *Virtual Voyages, Cinema and Travel*. Durham, NC: Duke University Press, 2006.

Sack, Robert. *Human Territoriality: Its Theory and History*. Cambridge: Cambridge University Press, 1986.

Sauerländer, Willibald. *Gothic Sculpture in France, 1140–1270*. New York: H. N. Abrams, 1973.

Scafi, Alessandro. *Mapping Paradise. A History of Heaven on Earth*. London: The British Library, 2006.

Schader, Jörg. "Jakobspilgern: Eine teilnehmende Beobachtung am Camino Francés." PhD diss., Karl-Franzens-University Graz, 2009.

Scheyvens, Regina. "Backpacker Tourism and Third World Development." *Annals of Tourism Research* 29 (2002): 144–64.

Schindl, Thomas. *Räume des Medialen: Zum spatial turn in Kulturwissenschaft und Medientheorie*. Boizenburg: Werner Hälsbusch 2007.

Schlittler, Anna-Brigitte and Katharina Tietze. *Kleider in Räumen.* Winterthur: Alataverlag, 2009.

Schmitt, Carl. *Der Nomos der Erde in Völkerrecht des Jus Publicum Europaeum.* Berlin: Duncker u. Humblot, 1974.

Schreiber, W. L. *Handbuch der Holz- und* Metallschnitte des XV. Jahrhunderts Volume IV. Leipzig: Karl W. Hiersamann, 1927.

Schroer, Markus. *Räume, Orte, Grenzen: Auf dem Weg zu einer Soziologie des Raums.* Frankfurt: Suhrkamp, 2006.

Schubert, Gabriella. *Kleidung als Zeichen: Kopfbedeckungen im Donau-Balkan-Raum.* Wiesbaden: Harrassowitz, 1993.

Schwartz, Barry. *Vertical Classification: A Study in Structuralism and the Sociology of Knowledge.* Chicago: University of Chicago Press, 1981.

Scott, James. "On Earth as in Heaven: The Apocalyptic Vision of World Geography from Urzeit to Endzeit According to the Book of Jubilees." In *Geography and Ethnography: Perceptions of the World in Pre-Modern Societies*, edited by Kurt A. Raaflaub and Richard J. A. Talbert, 182–96. Chichester, UK: Wiley-Blackwell, 2010.

Segal, Alan F. *Life after Death: A History of the Afterlife in Western Religion.* New York: Doubleday, 2004.

Seidel, J. Jürgen. *Die Anfänge des Pietismus in Graubünden.* Zürich: Chronos, 2001.

Serres, Michel. *Les Origines de la Géometrie.* Paris: Flammarion, 1993.

Sheldrake, Philip. *Spaces for the Sacred: Place, Memory, and Identity.* Baltimore, MD: Johns Hopkins University Press, 2001.

Shephard, Robert. "UNESCO and the Politics of Cultural Heritage in Tibet." *Journal of Contemporary Asia* 36.2 (2006): 243–57.

Shulman, David and Guy D. Stroumsa. "Introduction: Person, Passages, and Shifting Cultural Space." In *Self and Self-Transformation in the History of Religions*, edited by David Shulman and Guy S. Stroumsa, 3–18. Oxford: Oxford University Press, 2002.

Simon, Hermann. *The New Synagogue: Berlin: Past—Present—Future*, 3rd edition. Berlin: Edition Hentrich Berlin, 1999.

Simpson, von, Otto. *The Gothic Cathedral: Origins of Gothic Architecture and the Medieval Concept of Order.* Princeton, NJ: Princeton University Press, 1988.

Simpson, Moira. *Making Representations: Museums in the Post-Colonial Era.* London: Routledge, 1996.

Smith, Jonathan Z. *Map Is Not Territory.* Chicago: University of Chicago Press, 1978.

Smith, Murray. *Engaging Characters: Fiction, Emotion, and the Cinema.* Oxford: Clarendon Press, 1995.

———. "Imagining from the Inside." In *Film Theory and Philosophy*, edited by Richard Allen and Murray Smith, 412–30. New York: Oxford University Press, 1997.

Snodgrass, Adrian. *Architecture, Time and Eternity.* New Delhi: Aditya Prakashan, 1990.

Soja, Edward. *Thirdspace. Journeys to Los Angeles and Other Real and Imagined Places.* Cambridge: Blackwell, 1996.

Spitzer, Laura. "The Cult of the Virgin and Gothic Sculpture: Evaluating Opposition in the Chartres West Facade Capital Frieze." *Gesta* 33.2 (1994): 132–50.

Spreitzhofer, Günter. "Drifting und Travelling: 30 Jahre Rucksacktourismus in Südostasien: (K)ein Beitrag zur Entwicklung? Eine Bilanz." In *Tourismus in der "Dritten Welt". Zur Diskussion einer Entwicklungsperspektive*, edited by Herbert Baumhackl, Gabriele Habinger, Franz Kolland, and Kurt Luger, 90–120. Wien: Promedia, 2006.

Stausberg, Michael. *Religion im Modernen Tourismus.* Berlin: Insel, 2010.

Stobart, Jon, Andrew Hann, and Victoria Morgan, *Spaces of Consumption, Leisure and Shopping in the English Town, c. 1680–1830.* New York: Routledge, 2007.

Stoddard, Whitney S. *Art and Architecture in Medieval France: Medieval Architecture, Sculpture, Stained Glass, Manuscripts; The Art of the Church Treasuries.* Boulder, CO: Westview Press, 1972.

Stokstad, Marilyn. *Medieval Art.* 2nd edition. Boulder, CO: Westview, 2004.

Stolz, Fritz. "Ausbreitungsstrategien und Universalisierungstendenzen in der europäischen Religionsgeschichte." In *Europa verstehen*, edited by Ingolf Dalfert, Hans J. Luibl, and Hans Weder, 39–63 Zürich: TVZ, 1997.

———. "Effekt und Kommunikation" (1998). In *Religion und RekonstruktionI*, edited by Daria Pezzoli-Olgiati, 117–38. Göttingen: Vandenhoeck and Ruprecht, 2004.

———. *Grundzüge der Religionswissenschaft.* Göttingen: Vandenhoeck & Ruprecht, 2001.

———. *Weltbilder der Religionen: Kultur und Natur. Diesseits und Jenseits. Kontrollierbares und Unkontrollierbares.* Zürich: Pano, 2001.

Stone, Gregory. "Appearance and the Self." In *Dress and Identity*, edited by Mary E. Roach-Higgins, Joanne B. Eicher, and Kim K.P. Johnson, 19–39. New York: Fairchild Publications, 1995.

Strachan, James. *Early Bible Illustrations: A Short Study Based on Some Fifteenth and Early Sixteenth Century Printed Texts.* Cambridge: Cambridge University Press, 1957.

Stump, Roger W. *The Geography of Religion: Faith, Place, and Space.* Lanham, MD: Rowman and Littlefield, 2008.

Sutcliffe, William. *Are You Experienced?* London: Penguin, 1997.

Tagliaferri, Roberto. *Saggi di architettura e di iconografia dello spazio sacro.* Padova: Edizioni Messaggero, 2011.

Taylor, Lou. *The Study of Dress History.* New York: Manchester University Press, 2002.

Thiel, Erika. *Geschichte des Kostüms: Die europäische Mode von den Anfängen bis zur Gegenwart.* Berlin: Henschel, 2000.

Thompson, Kristin, David Bordwell, and Janet Staiger. *The Classical Hollywood Cinema: Film Style and Mode of Production to 1960.* New York: Columbia University Press, 1985.

Thompson, Leonard L. "Mapping an Apocalyptic World." In *Sacred Places and Profane Spaces: Essays on the Geographics of Judaism, Christianity, and Islam*, edited by Jamie Scott and Paul Simpson-Housley, 115–30. Westport, CT: Greenwood Press, 1991.

Todorov, Tzvetan. "Tribunals, Apologies, Reparations, and the Search for Justine: In Search of Lost Crime." *The New Republic* (January 2001): 9–10.

Tore, Ahlbäck, and Björn Dahla, eds. *Pilgrimage Today*. Åbo: Donner Institute for Research in Religious and Cultural History, 2010.

Trevithick, Alan. *The Revival of Buddhist Pilgrimage*. Dehli: Motilal Banarsidass, 2006.

The Tree of Life, directed by Terrence Malik. 2011. Los Angeles, CA: Fox Searchlight.

Turner, Victor. *The Ritual Process: Structure and Anti-Structure*. London: Routledge & K. Paul, 1969.

Turner, Victor and Edith L. B. Turner. *Image and Pilgrimage in Christian Culture*. New York: Columbia University Press, 1978.

Tweed, Thomas A. "Space." *Material Religion* 7.1 (2011): 116–23.

Uglow, Jennifer. *Hogarth: A Life and a World*. London: Faber and Faber, 1997.

Varela, Consuelo, ed. *Cristóbal Colón: Textos y documentos completes*. Madrid: Alianza Editorial, 1982.

Venter, Pieter M. "Daniel 9: A Penitential Prayer in Apocalyptic Garb." In *Seeking the Favor of God: Volume 2: The Development of Penitential Prayer in Second Temple Judaism*, edited by Mark J. Boda, Daniel K. Falk, and Rodney A. Werline, 33–50. Atlanta: Society of Biblical Literature, 2007.

Verkaaik, Oskar. "In Search of the Urban Bedou: The New Mosque of Granada." Unpublished paper.

Vernant, Jean-Pierre. *Mythe et pensée chez les Grecs: Études de psychologie historique*. Paris: Maspero, 1966.

Vorgrimler, Herbert. *Geschichte der Hölle*. München: Wilhelm Fink, 1994.

Watsuji, Tetsuro. *Fudo–Wind und Erde: der Zusammenhang zwischen Klima und Kultur*. Darmstadt: Wissenschaft Buchgesellschaft 1997.

Wechsler, Judith, Patricia Coronel, Michael Coronel, Sheila S. Blair, Jonathan M. Bloom, and E. Michael Whittington. "Caricature." In *Grove Art Online*. Oxford University Press, 2007–. Accessed September 7, 2010, http://oxfordartonline.com/subscriber/article/grove/art/T014063.

Weder, Hans. *Kosmologie und Kreativität*. Leipzig: Evangelische Verlagsanstalt, 1999.

Weinberg, Steven. *The First Three Minutes*. New York: Basic Books, 1977.

Welk, Peter. "The Beaten Track: Anti-Tourism as an Element of Backpacker Identity Construction." in *The Global Nomad: Backpacker Travel in Theory and Practice,* edited by Greg Richards and Julie Wilson, 77–91. Clevedon, UK: Channel View Publications, 2009.

———. "The Lonely Planet Myth: 'Backpacker Bible' and 'Travel Survival Kit.'" In *Backpacker Tourism: Concepts and Profiles*, edited by Kevin Hannam and Irena Ateljevic, 82–94. Clevedon, UK: Channel View Publications, 2008.

Westerhausen, Klaus. *Beyond the Beach: An Ethnography of Modern Travellers in Asia*. Bangkok: White Lotus Press, 2002.

White, Joanna. "Marks of Transgression: The Tattooing of Europeans in the Pacific Islands." In *Tattoo. Bodies, Art and Exchange in the Pacific and the West*, edited by Nicholas Thomas, Anna Cole, and Bronwen Douglas, 72–89. London: Reaktion Books, 2005.

Wilber, Ken. *No Boundary: Eastern and Western Approaches to Personal Growth*. Los Angeles: Center Publications, 1979.

Williams, Jane Welch. *Bread, Wine and Money: The Windows of the Trades at Chartres Cathedral*. Chicago: University of Chicago Press, 1993.

Williams, China, Celeste Brash, Andrew Burke, Shawn Low, Brandon Presser, Nick Ray, Daniel Robinson, Ryan Van Berkmoes, Richard Waters, Greg Bloom. *Southeast Asia on a Shoestring. Big Trips on Small Budgets*, 15th edition. London: Lonely Planet Publications, 2010.

Williams, Linda. "Film Bodies: Gender, Genre, and Excess." *Film Quarterly* 44.4 (1991): 2–13.

Williamson, Paul. *Gothic Sculpture, 1140–1300*. New Haven, CT: Yale University Press, 1998.

Wilson, Julie and Greg Richards. "Backpacker Icons: Influential Literary 'Nomads' in the Formation of Backpacker Identities." In *The Global Nomad: Backpacker Travel in Theory and Practice*, edited by Greg Richards and Julie Wilson, 123–45. Clevedon, UK: Channel View Publications, 2009.

Woiwoode, Christoph and Wolfgang Schotz, "Urban Development and Spirituality Trialog." *Zeitschrift für das Planen und Bauen in der Dritten Welt* 107.4 (2010): 18–26.

Würtz, Stefanie. and Roland Eckert. "Aspekte modischer Kommunikation." In *Inszenierungsgesellschaft. Ein einführendes Handbuchl*, edited by Herbert Willems and Martin Jurga, 177–91. Wiesbaden: Westdeutscher, 1998.

Young, Iris Marion. "The Ideal of Community and the Politics of Difference." In *Feminism/Post-Modernism*. edited by L. Nicholson, 300–23. London: Routledge, 1990.

Zuse, Konrad. *Rechnender Raum*. Braunschweig: Vieweg, 1969.

Contributors

Arnold O. Benz is professor emeritus at the Institute for Astronomy of ETH Zurich where he investigates star formations, solar physics, and plasma astrophysics. He is the author of the book *The Future of the Universe—Chance, Chaos, God?*

Pierre Bühler is professor for Systematic Theology and director of the Institute for Hermeneutics and Philosophy of Religion at University of Zurich. His main research fields are theological and philosophical hermeneutics; Luther- and Kierkegaard-research; theology and literature; hermeneutics and ethics.

Carla Danani is professor of Political Philosophy at the University of Macerata and professor of Philosophy of Dwelling at the Politecnico di Milano. Her research has been primarily in the field of Hermeneutics, Ethics, and Political Philosophy, with particular reference to the relation with space-territory and the strategies to implement a decision-making process through social participation.

Franco Farinelli is professor of Geography at Bologna University (Italy). His research includes the history of geographical thought, space and spatial studies, mapping and relationships between geography and philosophy.

Carl Fingerhuth's preoccupation has always been the town—as an archeologist in Egypt, with an office for town planning in Zürich, as the state architect in Basel, as a traveling expert in Europe, Asia, Africa, and America and as a teacher in universities in Switzerland, France, Italy, Germany, and the United States. He is an honorary professor at the Technical University of D-Darmstadt. His main interest is the quest for the "Gestalt" of the town beyond the modern, recently documented in the book *Learning from China—The Tao of the City.*

Mark K. George is associate professor of Hebrew Bible at the Iliff School of Theology in Denver, Colorado. His research includes the study of space and place, the Hebrew Bible and ancient Near East, aniconism, ritual, and critical theory and the Bible.

Monika Glavac has been working since 2006 for her PhD in the Study of Religion at the University of Zurich. The subject of her thesis is the relation between caricature and religion with focus on caricatures from the European history of religion representing Muslims. The book *Der 'Fremde' in der europäischen Karikatur* was published 2012 by Vandenhoeck and Ruprecht.

Anna-Katharina Höpflinger is a postdoctoral researcher in the Study of Religion at the University of Zurich. Her research interests focus on media and religion, clothing and religion, gender and religion, the dragon-slayer motif, religions in the ancient world, and heavy metal and religion.

Jacob N. Kinnard is professor of Comparative Religions, Iliff School of Theology and teaches courses on the religious traditions of India, comparative religion, and theoretical and methodological issues in the study of religion. Among the topics he writes about are Buddhist art and ritual, temples and pilgrimage places, and the ways in which religious communities understand and negotiate religiously significant places.

Remo H. Largo is professor emeritus of Children's Hospital Zurich, Switzerland. His main research areas are child growth and development and developmental disturbances in childhood and adolescence.

Marie-Therese Mäder, scholar in studies of religion and film studies, is currently a teaching and research fellow at the Center of Religion, Economy and Politics (in collaboration with the universities of Basel, Lausanne, Luzern, and the Collegium Helveticum) at the University of Zurich. Her research focus lies in audiovisual media and their interactions with religion, culture, and society. She is currently working with documentarizing media and the reconstruction of religion.

Christl M. Maier is professor of Old Testament at Philipps University Marburg, Germany. Her research includes the prophetic writings and wisdom literature of the Hebrew Bible, feminist hermeneutics, and spatial studies.

Tommi Mendel is a PhD candidate at the Institute of Religious Studies at the University of Zurich, Switzerland. His research includes pilgrimages and backpacking, audiovisual methods, ethnographic film, film and religion as well as Japanese religion and culture.

Daria Pezzoli-Olgiati is a professor of the Study of Religion and director of the Center for Religion, Economy, and Politics at the University of Zurich, Switzerland. Her interests focus on visible and material religion, space- and gender-critical approaches to religion, ancient and European history of religion, apocalypticism and methods, and theories of the Study of Religion.

S. Brent Plate is visiting associate professor in Religious Studies at Hamilton College. He is cofounder and managing editor of the award-winning *Material Religion: The Journal of Objects, Art, and Belief.* Among his other books, *A History of Religion in 5½ Objects* was published in 2014 by Beacon Press.

Katherine Rousseau is a PhD candidate in Religious Studies at Iliff School of Theology and the University of Denver in Denver, Colorado, and the assistant director for International Affairs at the University of Colorado Boulder. Her research interests include spatial studies, pilgrimage, and the history of art and visual culture.

Index

Printed and bound in the United States of America